Polyploidy: Recent Trends and Future Perspectives

Tanvir-Ul-Hassan Dar • Reiaz-Ul Rehman

Polyploidy: Recent Trends and Future Perspectives

Springer

Tanvir-Ul-Hassan Dar
Department of Biosciences and Biotechnology
Baba Ghulam Shah Badshah University
Rajouri, Jammu and Kashmir, India

Reiaz-Ul Rehman
Department of Bioresources
University of Kashmir
Srinagar, Jammu and Kashmir, India

ISBN 978-81-322-3770-9 ISBN 978-81-322-3772-3 (eBook)
https://doi.org/10.1007/978-81-322-3772-3

Library of Congress Control Number: 2017956126

Printed on acid-free paper

This Springer imprint is published by Springer Nature
The registered company is Springer (India) Private Ltd.
The registered company address is: 7th Floor, Vijaya Building, 17, Barakhamba Road, New Delhi 110 001, India

Contents

About the Authors

Dr. Tanvir-Ul-Hassan Dar, PhD, is presently working as an assistant professor at the School of Biosciences and Biotechnology, Baba Ghulam Shah Badshah University, Rajouri (Jammu and Kashmir). He completed his master's degree in biotechnology at the University of Kashmir, Srinagar, and obtained his MPhil and PhD from the University of Delhi. Subsequently, he had one year of postdoctorate research training at the National Institute of Plant Genome Research (NIPGR), New Delhi. Dr. Dar has 5 years of teaching experience at the postgraduate level and is currently involved in supervision of MPhil and PhD students. With specialization in polyploidy, he has acquired the necessary training and experience to carry out research in this field of biological sciences. He has published more than 15 research papers in peer-reviewed national and international journals; attended numerous seminars, conferences, and workshops; and is a member of several important academic and scientific bodies.

Reiaz-Ul Rehman, PhD, is currently an assistant professor at the Department of Bioresources, School of Biological Sciences, University of Kashmir, and is actively involved in teaching master's and integrated PhD students. He earned his master's and PhD in plant biology at Hamdard University, New Delhi, and has worked as a postdoctoral research fellow at the University of Louisville (Kentucky, USA) and the University of Salento (Lecce, Italy). He is a recipient of Marie Curie International Incoming Fellowship (IIF) (Lisbon, Portugal). His research interests include the characterization of the physiological and metabolic processes in underutilized crops challenged with stress. He has published 40 research articles and one edited book.

Introduction to Polyploidy

1

1.1 Introduction

Polyploidy refers to the presence of more than two genomes per somatic cell. Generally, the polyploid organism would have multiple sets of chromosomes or either the combination of chromosome sets found in same species or a closely related diploid species. Polyploid organisms can arise spontaneously by the chromosome duplication of somatic cells, or they can arise during meiosis due to non-disjunction of homolog chromosomes giving rise to diploid gametes (Ramsey and Schemske 2002). Polyploidy can also be induced artificially in the laboratory by drug treatment of the cells of an organism which inhibits the cell division, e.g. colchicine. In nature the polyploidy exists in organisms in either all or some somatic cells. The prevalent occurrence of polyploids suggests that polyploidy confers some selective advantages over diploidy. Some of the advantages can be enhanced vigour, increased heterozygosity, novel variation and production of new genotypes and allelic sub-functionalization (Salmon et al. 2005; Udall and Wendel 2006; Abbott et al. 2007; Anssour et al. 2009; Dar et al. 2013). The ancestral condition of any eukaryote is now suspected to have gone through polyploidization and re-polyploidization (Blanc et al. 2003; Bowers et al. 2003; Tuskan et al. 2006; Jaillon et al. 2007). The return of many genes back to single copy with disomic chromosomal inheritance and, in some cases, to smaller genome size and reduced chromosome numbers can occur through diploidization. Diploidization mechanisms include events such as chromosomal translocations, deletion of repetitive sequences and gene silencing (Pontes et al. 2004; Han et al. 2005; Gaeta et al. 2007; Mandáková et al. 2010).

Being one of the best known mechanisms for rapid mode of speciation, polyploidy has an important role in the formation of new species (reviewed in Otto and Whitton 2000; Doyle et al. 2008; Leitch and Leitch 2008). Hybridization and sympatric speciation through polyploidy are well-established phenomena in plants and have played significant roles in the evolution and 'macro-diversification' of flowering plants (Soltis and Soltis 2009). Fawcett et al. (2009) suggest that polyploidy helped numerous plant lineages survive the Cretaceous-Tertiary (K-T) mass

T.-Ul.-H. Dar, R.-Ul. Rehman, *Polyploidy: Recent Trends and Future Perspectives*, https://doi.org/10.1007/978-81-322-3772-3_1

extinction. It has also been found that duplicate genes increase gene expression diversity in closely related species and allopolyploids, thereby facilitating morphological and adaptive evolution (Wang et al. 2004; Adams and Wendel 2005; Ha et al. 2009); Vamosi and Dickinson (2006) provided an analysis linking polyploidy and species richness in Rosaceae; based on comparisons of diversification rates, Soltis and Soltis (2009) suggest that genome doubling may have led to a dramatic increase in species richness in several angiosperm lineages, including Poaceae, Solanaceae, Fabaceae and Brassicaceae. These and other such findings negate the long-held belief that polyploids are commonly 'evolutionary dead ends' (for review see Soltis and Soltis 1993). This prompts one to believe that, as a process of speciation, polyploidy holds great promise for future to make up for the unprecedented loss of biodiversity being faced all over the world.

Initially, polyploids gained attention because of their unique cytogenetics and reproductive isolation from diploids (Blakeslee 1921; Jorgensen 1928). Subsequently, it was recognized that polyploids also exhibit distinctive phenotypic traits, pertaining to morphology, anatomy, physiology, reproductive characters, phenology and overall life history, which differentiate neo-polyploids from their progenitor diploids (Pal and Khoshoo 1977; Hahn et al. 1990). 'Gigas' characteristics, such as sturdier foliage, thicker stems and enlarged reproductive structures, which are typically less numerous in the diploid progenitors, are some of the commoncharacteristics of neo-polyploids. Compared to their diploid ancestors, neo-polyploids have been reported to often show increase in the size of individual cells, reduced branching, variations in the contents of various substances, slower growth and development, relatively later and longer flowering and reduction of pollen and seed fertility (see Stebbins 1950; Ramsey and Schemske 2002).

The above-said genomic changes induced by polyploidy are thought to influence the transcriptome, metabolome and proteome, which can change the phenotype and ecology of the individuals (Leitch and Leitch 2008). Song et al. (1995) reported changes in genomic organization of the synthetic allotetraploid of *Brassica* sp. and detected nonadditive inheritance of genomic fragments. Both the loss of parental genome fragments and the gain of new fragments were observed. Allotetraploids of wheat have also displayed genomic changes immediately after hybrid formation (Feldman et al. 1997; Ozkan et al. 2001; Shaked et al. 2001; Town et al. 2006). These findings, besides explicitly negating the long-standing perception that polyploids are evolutionary dead ends, not dynamic (Wagner 1970), suggest that polyploidy is almost a universal phenomenon.

Polyploids are of four main types, viz. auto-, allo-, segmental and auto-allopolyploids (Stebbins 1947). Autopolyploid organisms have similar genome which arises by genome duplication thus resulting in multiple similar sets of chromosomes (Stebbins 1947; Lewis 1980). In allopolyploid organisms two or more divergent chromosome sets get merged (Stebbins 1947; Grant 1975). The segmental allopolyploids are formed from the two or more partially different genomes giving rise to bivalents and multivalents during chromosome pairing. In auto-allopolyploids, both auto- and allopolyploids get combined (Table 1.1).

Table 1.1 Classification of polyploidy

Type	Description	Genomic constitution	Example
Autopolyploidy	A polyploid having multiple sets of identical genomes	AAAA	*Solanum tuberosum, Zea perennis*
Allopolyploidy	A polyploid having multiple sets of different genomes	AABB	*Triticum aestivum, Gossypium hirsutum*
Segmental allopolyploids	A polyploid having multiple sets of similar but not identical genomes	$A_1A_1A_2A_2$	*Aesculus carnea, Delphinium gypsophilum*
Autoallopolyploids	A polyploid produced by the combination of autopolyploidy and allopolyploidy	AABBBB	*Helianthus tuberosus*

1.2 Historical Perspective

Etymologically, polyploidy is composed of two words: *poly*, Greek for many/more than two, and *ploideus*, ending of cytological terms relating to the number of sets of chromosomes, derived from Greek *idios* (private, peculiar, individual) by way of Weismann's term *Id*, introduced in German in 1893 to designate the hereditary germplasm, and Strasburger's term haploid (single id, gametophyte) and diploid (double id, sporophyte), introduced in German in 1905 (Stearns 1992). The phenomenon is one of the earliest cytogenetic characteristics to be studied in greater detail (Stebbins 1950). Probably for the first time, *Oenothera gigas* De Vries was recognized to be a tetraploid mutant of *O. lamarckiana* (Lutz 1907; Gates 1909). Kuwada (1911) wrote on an early interest in putative chromosome duplication in maize (*Zea mays*). Winkler (1916) is said to have introduced the term polyploidy, when he artificially made a tetraploid form of *Solanumnigrum*, perhaps the first laboratory production of a polyploid type. Meanwhile, multiples of a basic chromosome number were revealed in many plant genera, such as *Drosera* (Rosenberg 1909), *Dahlia* (Ishikawa 1911) and *Chrysanthemum* (Tahara 1915). Digby (1912) found that the fertile *Primula kewensis* (a sterile interspecific hybrid *P. verticillata* × *floribunda*) could occur by doubling its chromosome number. Winge (1917) compared chromosome numbers of related species in *Chenopodium* and *Chrysanthemum* and found that in both cases, they are multiples of a common basic number. This led him propose a fruitful hypothesis that hybridization followed by polyploidy is a method for the origin of species. Winge's hypothesis was soon confirmed by the artificial production of several other species through interspecific hybridizations, notable among them being the classical combination of radish and cabbage, *Raphanobrassica* (Karpechenko 1927, 1928), and the evidence that *Spartina townsendii* is derived from the cross of *S. alterniflora* × *stricta* (Huskins 1931a,b). Simultaneously, interspecific hybridization was noticed in some other important genera, such as *Triticum* (Sakamura 1918; Sax 1922), *Rosa* (Tackholm

1922) and *Nicotiana* (Clausen and Goodspeed 1925). Muntzing (1936) and Darlington (1937) provided early reviews of polyploidy in plants.

It was the discovery of the colchicine technique of doubling chromosome number by Dustin et al. (1937) and its practical use by Blakeslee and Avery (1937) that provided a vital tool for experimental studies of polyploidy (Eigsti 1947). Meanwhile, it became known that many of our most valuable crop plants, such as wheat, oats, cotton, tobacco, potato, banana, coffee and sugar cane, are polyploids (reviewed in Clausen et al. 1945; Löve and Löve 1949; Stebbins 1940, 1947, 1950). The importance of polyploidy in synthesizing new types by the plant breeder was recognized in the studies of Crane and Darlington (1927), Crane (1940a), Thomas (1940a, b), Yarnell (1936), Darrow et al. (1944) and Darrow and Camp (1945). With the publication of the famous book *Variation and Evolution in Plants* by Stebbins (1950), and his important work (1971), studies on polyploidy continued with renewed vigour and much attention has been paid to the phenomenon throughout the world (Lewis 1980). Now it is known that polyploidy is extremely common, a characteristic feature, in all the major groups of plants except the gymnosperms (Stace 2000). The process of genome duplication has been found to have also recurred in plant lineages during their evolutionary history. As a result, the relationships of polyploids with each other and with their diploid progenitors are complex, making the application of species concepts to them difficult (Rieseberg and Willis 2007; Soltis et al. 2007).

1.2.1 Evolution of Important Crops

1.2.1.1 Origin of Tetraploid *Brassica* spp.

The creation of allotetraploid hybrid of *Brassica juncea* (AABB, $2n = 36$); *Brassica napus* (AACC, $2n = 38$) and *Brassica carinata* (BBCC, $2n = 34$) has taken place by allopolyploidization events during last 10,000 years (Nagaharu 1935). Cultivated *Brassica* is comprised of six species, three diploids and three allopolyploids derived from three diploid species. The genetic relationship between these six species has been confirmed by artificial synthesis as represented in triangle of U (U, 1935). In the U triangle, it is shown that *Brassica juncea* ($2n = 36$, AABB) is derived from that *B. nigra* ($2n = 16$, BB) and *B. rapa* ($2n = 20$, AA). *Brassica napus* ($2n = 38$, AACC) is derived from *B. oleracea* ($2n = 18$, CC) and *B. rapa* ($2n = 34$, BBCC). *Brassica carinata* ($2n = 34$, BBCC) is derived from *B. nigra* ($2n = 16$, BB) and *B. oleracea* ($2n = 18$, CC) (Fig. 1.1).

1.2.1.2 Origin of Bread Wheat

Two sequential allopolyploidization events are involved in the origin of bread wheat. First, allopolyploidization between two diploid species *Triticum urartu* (A genome donar) and goatgrass species (undiscovered or extinct) closely related to *Aegilops speltoides* (B genome donar) led to the origin of allotetraploid wheat 0.36–0.5 million years ago. Second, allopolyploidization that took place around 10,000 years ago between cultivated *Triticum turgidum* ssp. *durum* and goat grass, *Aegilops tauschii* (D genome donar), led to the formation of bread wheat (Zhang et al. 2014) (Fig. 1.2).

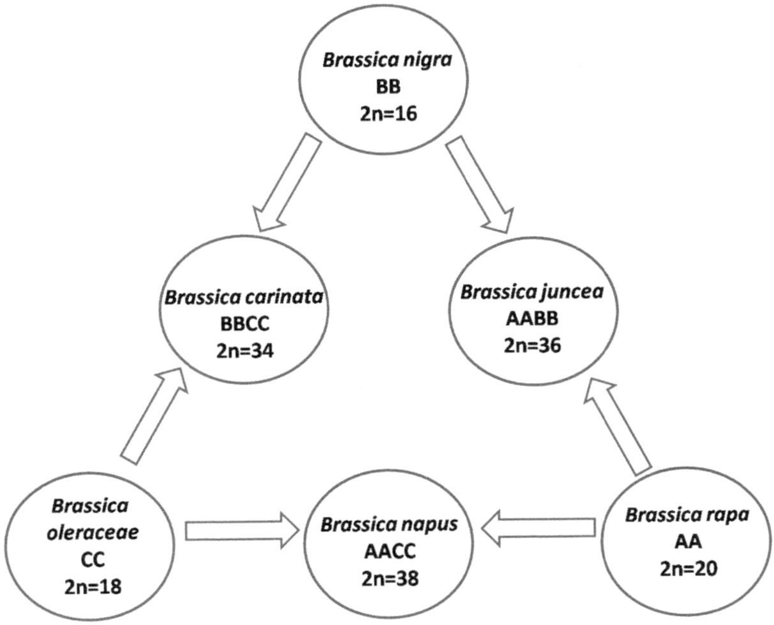

Fig. 1.1 U-Triangle depicting the origin of three tetraploid cultivated *Brassica* spp. from three diploid progenitor species

1.2.1.3 Origin of Gossypium

Origin of *Gossypium* has a long history. The genus contains around 50 diploid species ($2n = 2X = 26$) and five allotetraploid ($2n = 4X = 52$) species. Much of the attention has been paid to four domesticated species including the new world allopolyploid diploids (*G. barbadense* and *G. hirsutum*) and the old world diploids (*G. herbaceum* and *G. arboretum*). Based on molecular data including cpDNA and 5 s ribosomal analysis, *Gossypium* species genealogical lineages are largely congruent with genome designations and geographical distributions (Wendel and Albert 1992; Cronn et al. 1996). Each genome group leads to a single natural lineage. In most cases these lineages are also geographically cohesive (Fig. 1.3). There are three major lineages of diploid species, corresponding to three continents. These are Australia (C, G and K genomes), the Americas (D genome) and Africa/Arabia (A, B, E and F genomes). The divergence among major groups appears due to divergence of DNA sequences, having been established by approximately 11.5 million base pairs. New world tetraploid (AD genome) arises due to inter-genomic hybridization between new world cotton (D genome) and African-Asian (A genome) species.

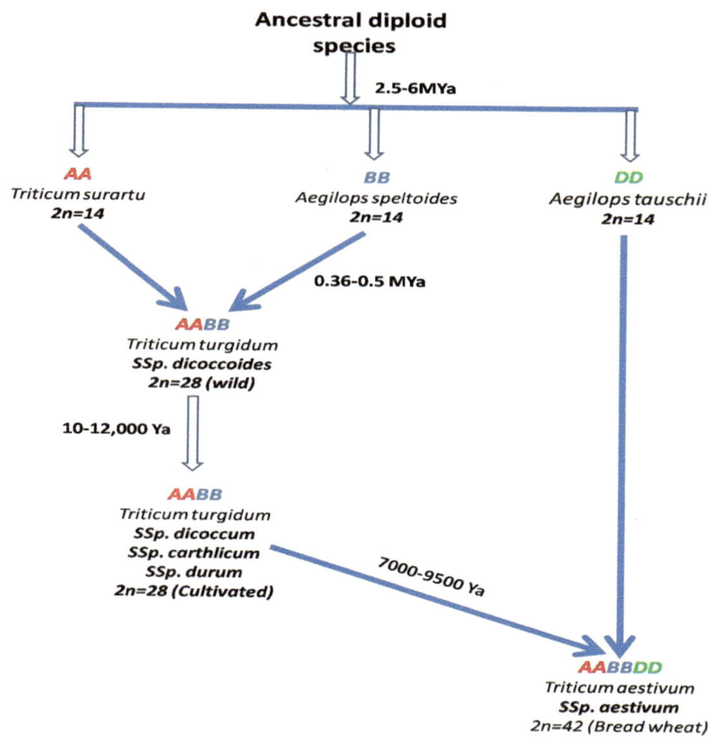

Fig. 1.2 Evolution of bread wheat from its diploid progenitors

1.3 Recent Polyploidization

The effect of genome doubling on polyploidization can be best studied in polyploids with recent ancestry. It has been reported that only a few polyploid species have arisen from the past 150 years. These include *Spartina anglica*, *Cardamine schulzii*, *Senecio cambrensis*, *Senecio eboracensis* and *Tragopogon mirus* and *T. miscellus* (reviewed in Soltis et al. 2004).

Tragopogan genus belongs to the family Asteraceae and is composed of approximately 150 species, all native to Europe. Most of these species are diploids ($2n = 12$), but some polyploid species have also been reported. The introduction of diploid *Tragopogan* (*Tragopogon dubius* Scop., *Tragopogan pratensis* L. and *Tragopogan porryfolius* L.) to North America from Europe led to the formation of allopolyploids (*Tragopogan mirus* Ownbey and *Tragopogan miscellus* Ownbey) in the last 80 years, and thus the latter are ideal candidates for studying recent polyploidization events. Morphological and cytological data (Ownbey 1950) revealed that *T. mirus* and *T. miscellus* are allotetraploid ($2n = 24$) and their diploid ($2n = 12$) parents are *T. dubius* and *T. porryfolius* and *T. dubius* and *T. pratensis*, respectively (Fig. 1.4). The analysis of *Tragopogan* polyploids at cytogetic and genomic levels has been studied by analysing rDNA ITS + ETS which indicated that parental diploids are phylogenetically divergent within *Tragopogan*genus (Soltis et al. 2004). Concerted

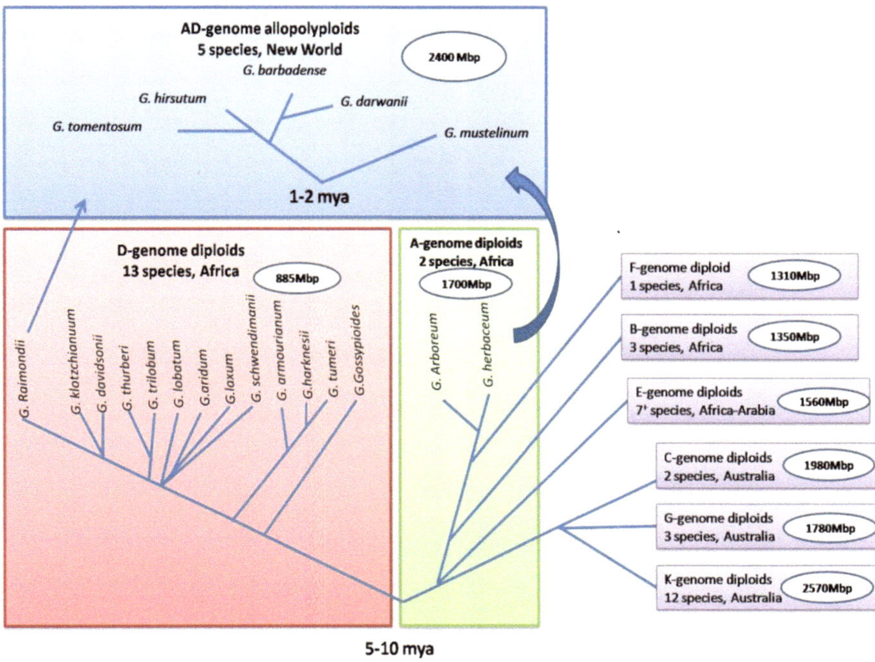

Fig. 1.3 Evolutionary history of Gossypium. Different diploid groups and their relationship are shown. Formation of allotetraploid Gossypium as a result of inter-genomic hybridization

evolution of rDNA was confirmed by using southern blotting and cloning experiments on tissues from old herbarium collections of *Tragopogan* and comparing them with recent samples of the genus. It is being said that the concerted evolution is an ongoing process which has not proceeded to completion in any of the polyploidy population studied so far. It has been seen that the rDNA repeats in diploid *Tragopogan dubius* are converted or lost in allopolyploids including those with independent origin. Molecular cytogenetic studies using sub-telomeric repeats, centromeric repeats and rDNA as probes helped in distinguishing all chromosomes among diploid progenitors of *Tragopogan* ($2n = 12$). It was seen that the diploid chromosome complex are additive in both above polyploids ($2n = 24$). However, no evidence of chromosomal rearrangement was seen in *Tragopogan mirus* and *Tragopogan miscellus* populations. The cDNA AFLP displayed gene expression differences among *Tragopogan miscellus* and its diploid parents and also between *Tragopogan miscellus* populations of reciprocal origin. It was observed that 5% of genes in allopolyploid populations were silenced and 4% of the genes exhibited novel expression in comparison to their parents. Somehow these changes represent maternal or paternal effects. It can be summed from the above that the multiple origins of polyploidy species affect the genetic variation patterns in addition to differential gene expression patterns, thus having a major role in long-term evolution of polyploids (Soltis et al. 2004).

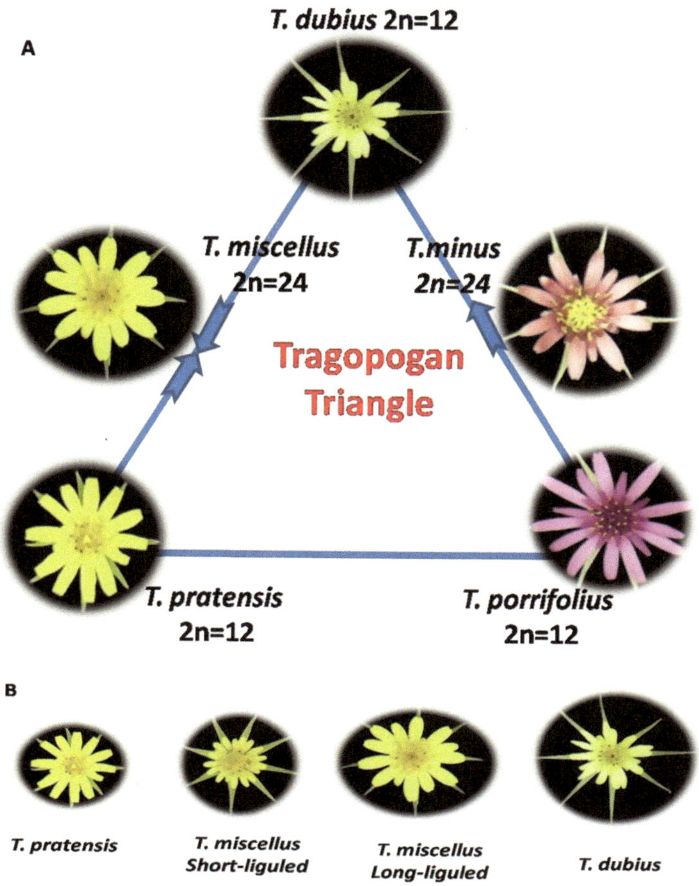

Fig. 1.4 (**a**)Tragopogon triangle showing recent origin of allotetraploid *T. miscellus* and *T. mirus*. Maternal parentages are indicated by arrows. *T. dubius* or *T. pratensis* are the maternal parents of *T. miscellus*. *T. porrifolius* is the maternal parent of *T. mirus* (**b**) *T. miscellus* populations of separate origin differ in morphology. Populations which had *T. pratensis* as maternal parent showed short ligules, whereas those had *T. dubius* as maternal parent showed long ligules (Modified from Soltis et al. 2004)

1.4 The Polyploidy Complex

Sometimes, some or all of the above-mentioned polyploid types may occur together with their diploid ancestors. If these diploids are so closely related to each other as to be able to produce segmental allopolyploids as well as autopolyploids, recombination of genomes between taxa will occur at the tetraploid and higher levels of ploidy. As such, the distinction seen between taxa at the diploid level becomes blurred at the tetraploid and higher levels, rendering recognition of distinct taxa impossible. This will result in varied pattern of a plant group, which Babcock and Stebbins (1938) called the *polyploid complex*. In such a complex, the diploid forms may be considered as a series of distinct pillars, supporting a large structure of intermediate polyploids, thus the name *polyploid pillar complex.*

There are numerous examples of polyploid complexes in various groups of plants. In polyploid complexes, the interrelationships between the different ploidy levels and between different taxa at each ploidy level are complicated (Stace 1989); thus they are difficult to deal with taxonomically. Such a plant group has usually diploids and at least one higher ploidy level. Whereas diploids can be distinguished, the taxa at higher ploidy levels show continuous genetic variation which blurs taxonomic demarcation. Stebbins (1971) divided polyploid complexes into five types (initial, young, mature, declining and relictual), which represent stages of their one-way evolution from a decreasing number of diploids to an increasing number of polyploids. Based on this, plant groups existing only as high polyploids can well be considered as approaching an endpoint in evolution (Stace 2000).

The term *permanent odd polyploidy* was used by Grant (1981) for a special type of genetic system, such as in a pentaploid agamospermous *Crepis* plant or a triploid viviparous *Allium* plant. By a permanent odd polyploid, Grant meant a plant which breeds true for an odd euploid condition by sexual means, transmitting one genome through the pollen and two or more genomes through the ovules. This system is found in *Rosa canina* group (Rosaceae) and *Leucopogon juniperinus* (Epacridaceae).

1.5 Polyploidy and Base Number

As per Henderson's *Dictionary of Biology* (13th ed. 2005), basic number or base number refers to 'the minimum haploid chromosome number occurring in a series of euploid species in a genus' or the 'chromosome number in gametes of diploid ancestor of a polyploid organism'. Chromosome base numbers have a pivotal role in assessing the extent of polyploidy, because it is the product of these two variables [(base number (x) and ploidy level (n)] that determines chromosome number; they are also frequently of great evolutionary significance and taxonomic value (Stace 2000). In the genera exhibiting a single base number, which expresses only in euploid manner, the proportion of polyploids is easy to determine, but it is not so in cases where the base number is more than one, or the aneuploidy is common. Based on assumption that the highest base number in angiosperms is 11 or 13, it is believed that 35–70% of these plants are polyploids and that the number of plant species decreases as we go higher up in the ploidy level. The highest level of ploidy is reported in the fern genus *Ophioglossum*, where the base number (x) = 15, and a race of *O. reticulatum* (with $2n = 1440$) is 96-ploid. Among angiosperms, the palm genus *Voanioala* ($2n = 596–606$) has been estimated as 38-ploid (Roser1994), while in the grass genus *Poa*, $x = 7$, so *P. litorosa* (with $2n = 266$) is 38-ploid.

1.6 Polyploidy and Aneuploidy

Aneuploidy refers to a condition of having more or fewer than an exact multiple of the haploid number of chromosomes. As against euploidy (duplication of complete chromosomal sets), the chromosome numbers in aneuploidy vary not in multiples of the basic set but in single or a few chromosomes only. For example, in *Vicia* $2n = 10$, 12, 14, 24, 28, etc., comprising aneuploids clustered around the diploid and tetraploid levels, while in *Crepis* $2n = 6$, 8, 10, 12, 14, 16, 18, 22, 24, 42, 44, 66, 88, etc., with no clear clustering, such a condition where numbers bear no clear relation to a particular series of ploidy levels has been called *dysploidy* (Stace 1989). Some authors hold that while aneuploidy involves also a loss of genes (Levin 2002), dysploidy is a change in haploid chromosome number without loss of genetic material, i.e. 'integral differences in chromosome numbers without appreciable genetic imbalance' (Strother and Brown 1988); sometimes, the two are also used interchangeably. One of the remarkable examples of a long and nearly continuous aneuploid series is provided by the genus *Carex*.

Aneuploid variation is also common and may involve loss or gain of a single chromosome, a single chromosome pair, or more than one pair. It can occur at the diploid or polyploid level. Whereas the gain of extra chromosomes (ascending aneuploidy) has better chances of tolerance at any ploidy level, the loss of some chromosomes from the basic set may even be lethal at the diploid level (Stace 1989). Aneuploid reduction (descending aneuploidy) at the polyploid level has been called *polyploid drop* (Darlington 1963; 1973). The loss of chromosomes in a polyploid can often be tolerated because of the presence of much duplicated genetic material in their nuclei and the buffering effect of multiple genomes (Grant 1981; Stace 1989).

References

Abbott RJ, Ireland HE, Rogers HJ (2007) Population decline despite high genetic diversity in the new allopolyploid species *Senecio cambrensis* (Asteraceae). Mol Ecol 16:1023–1033

Adams KL, Wendel JF (2005) Allele-specific, bidirectional silencing of an alcohol dehydrogenase gene in different organs of interspecific diploid cotton hybrids. Genetics 171:2139–2142

Anssour S, Krugel T, Sharbel TF, Saluz HP, Bonaventure G, Baldwin IT (2009) Phenotypic, genetic and genomic consequences of natural and synthetic polyploidization of *Nicotiana attenuata* and *Nicotiana obtusifolia*. Ann Bot 103:1207–1217

Babcock EB, Stebbins GL (1938) The American species of *crepis*. Publ Carnegie Inst Wash Publ 504:1–199

Blakeslee AF (1921) Types of mutations and their possible significance in evolution. Am Nat 55:254–267

Blakeslee AF, Avery AG (1937) Methods of inducing chromosome doubling in plants by treatment with colchicine. Science 86:408

Blanc G, Hokamp K, Wolfe KH (2003) A recent polyploidy superimposed on older large-scale duplications in the *Arabidopsis* genome. Genome Res 13:137–144

Bowers JE, Chapman BA, Rong J, Paterson AH (2003) Unravelling angiosperm genome evolution by phylogenetic analysis of chromosomal duplication events. Nature 422:433–438

Clausen RE, Goodspeed TH (1925) Interspecific hybridization in *Nicotiana*. II. A tetraploid *Glutinosa tabacum* hybrid, an experimental verification of Winge's hypothesis. Genetics 10:278–284

Clausen J, Keck DD, Hiesey WM (1945) Experimental studies on the nature of species. II. Plant evolution through amphiploidy and autoploidy, with examples from the Madiinae. Carnegie Institution Washington Publication No. 1945:564

Crane MB (1940) The origin and behavior of cultivated plants. In: Huxley J (ed) The new systematics, pp 529–547

Crane MB, Darlington CD (1927) The origin of new forms in *Rubus*. Genetica 9:241–277

Cronn RC, Zhao X, Paterson AH, Wendel JF (1996) Polymorphism and concerted evolution in a tandemly repeated gene family: 5S ribosomal DNA in diploid and allopolyploid cottons. J Mol Evol 42:685–705

Dar TH, Raina SN, Goel S (2013) Molecular analysis of genomic changes in synthetic autotetraploids *Phlox drummondii* Hook. Biol J Linn Soc 110:591–605

Darlington CD (1937) Recent advances in cytology. Blakiston, Philadelphia

Darlington CD (1963) Chromosome botany and the origins of cultivated plants, 2nd edn. Hafner, New York

Darlington CD (1973) Chromosome botany and the origin of cultivated plants. George Allen & Unwin Ltd, London

Darrow GM, Camp WH (1945) *Vaccinium* hybrids and the development of new horticultural material. Bull Torrey Bot Club 72:1–21

Darrow GM, Camp WH, Fischer HE, Dermen H (1944) Chromosome numbers in *Vaccinium* and related groups. Bull Torrey Bot Club 71:498–506

Digby L (1912) The cytology of *Primula kewensis* and of other related *Primula* hybrids. Ann Bot 26(2):357–388

Doyle JJ, Flagel LE, Paterson AH, Rapp RA, Soltis DE, Soltis PS, Wendel JF (2008) Evolutionary genetics of genome merger and doubling in plants. Annu Rev Genet 42:443–461

Dustin AP, Havas L, Lits F (1937) Réunion Assoc Anatomistes, Marseille

Eigsti OJ (1947) Colchicine bibliography. Lloydia 10:65–114

Fawcett JA, Maerea S, de PYV (2009) Plants with double genomes might have had a better chance to survive the Cretaceous–Tertiary extinction event. Proc Natl Acad Sci 106:5737–5743

Feldman M, Liu B, Segal G, Abbo S, Levy AA, Vega JM (1997) Rapid elimination of low-copy DNA sequences in polyploid Wheat: a possible mechanism for differentiation of homoeologous chromosomes. Genetics 147:1381–1387

Gaeta RT, Pires JC, Iniguez-Luy F, Leon E, Osborn TC (2007) Genomic changes in resynthesized *Brassica napus* and their effect on gene expression and phenotype. Plant Cell 19:3403–3417

Gates RR (1909) The stature and chromosomes of *Oenothera gigas*, De Vries. Arch Zellforsch 3:525–552

Grant V (1975) Genetics of flowering plants. Columbia University Press, New york

Grant V (1981) Plant speciation, 2nd edn. Columbia University Press, New York

Ha M, Lu J, Tian L, Ramachandran V, Kasschau Kristin D, Chapman EJ, Carrington JC, Chen X, Wang XJ, Chen ZJ (2009) Small RNAs serve as a genetic buffer against genomic shock in *Arabidopsis* interspecific hybrids and allopolyploids. Proc Natl Acad Sci 106:17835–17840

Hahn SK, Bai KV, Asiedu R (1990) Tetraploids, triploids, and 2*n* pollen from diploid interspecific crosses with cassava. Theor Appl Genet 79:433–439

Han F, Fedak G, Guo W, Liu B (2005) Rapid and repeatable elimination of a parental genome-specific DNA repeat (pGc1R-1a) in newly synthesized wheat allopolyploids. Genetics 170:1239–1245

Huskins CL (1931a) Origin of *Spartina townsendii*. Nature (London) 127:781

Huskins CL (1931b) The origin of *Spartina townsendii*. Genetica 12:531–538

Ishikawa M (1911) Cytologische studien von dahlien. Bot Mag Tokyo 25:1–8

Jaillon O, Aury JM, Noel B, Policriti A, Clepet C et al (2007) French-Italian public consort.grapevine genome charact. The grapevine genome sequence suggests ancestral hexaploidization in major angiosperm phyla. Nature 449:463–467

Jorgensen CA (1928) The experimental formation of heteroploid plants in the genus *Solanum*. J Genet 19:133–210

Karpechenko D (1927) Polyploid hybrids of *Raphanus sativus L*.X *Brassica oleracea L*. Bull Appl Bot 17(4):305–408. (In Russian, English summary, pp 398–408)

Karpechenko D (1928) Polyploid hybrids of *Raphanus aativua* L. X *Brassica oleracea* L. 2.indukt. Abstamm.- u. Vererb Lehre 48:1–85
Kuwada Y (1911) Meiosis in the pollen mother cells of Zea Mays L. Bot Mag Tokyo 25:163–181
Leitch AR, Leitch IJ (2008) Genomic plasticity and the diversity of polyploid plants. Science 320:481–483
Levin D (2002) The role of chromosomal change in plant evolution. Oxford Univ Press, Oxford
Lewis WH (1980) Polyploidy in species populations. In: Lewis WH (ed) Polyploidy: biological relevance. Plenum, New York, pp 103–144
Löve A, Löve D (1949) The geobotanical significance of polyploidy. I. Polyploidy and latitude. Port Acta Biol Ser A1949:273–352
Lutz AM (1907) A preliminary note on the chromosomes of *Oenothera Lamarckiana* and one of its mutants, *O. gigas*. Science 26:151–152
Mandáková T, Joly S, Krzywinski M, Mummenhoff K, Lysaka MA (2010) Fast diploidization in close mesopolyploid relatives of *Arabidopsis*. Plant Cell 22:2277–2290
Muntzing A (1936) The evolutionary significance of autopolyploidy. Hereditas 21:263–378
Nagaharu U (1935) Genome analysis in Brassica with special reference to the experimental formation of *B. napus* and peculiar mode of fertilization. Jpn J Bot 7:389–452
Otto SP, Whitton J (2000) Polyploidy incidence and evolution. Annu Rev Genet 34:401–437
Ownbey M (1950) Natural hybridization and amphiploidy in the genus Tragopogon. Am J Bot 37:487–499
Ozkan H, Levy AA, Feldman M (2001) Allopolyploidy induced rapid genome evolution in the Wheat (*Aegilops – Triticum*) group. Plant Cell 13:1735–1747
Pal M, Khoshoo TN (1977) Evolution and improvement of cultivated *Amaranths*. VIII. Induced autotetraploidy in grain types. ZPflanzenzuecht 78:135–148
Pontes O, Neves N, Silva M, Lewis MS, Madlung A, Comai L, Viegas W, Pikaard CS (2004) Chromosomal locus rearrangements are a rapid response to formation of the allotetraploid *Arabidopsis suecica* genome. Proc Natl Acad Sci U S A 101:8240–8245
Ramsey J, Schemske DW (2002) Neopolyploidy in flowering plants. Annu Rev Ecol Syst 33:589–639
Rieseberg LH, Willis JH (2007) Plant speciation. Science 317:910–914
Rosenberg O (1909) Cytologische und morphologische Studien an Drosera longifolia x rotundifolia. Kongl. Svenska Vetenskapsakad Handl 43:1–65
Sakamura T (1918) Kurze Mitteilung uber die Chromosomenzahlen und die Vcrwandtschtsverhaltnisse der *Triticum*. Artem. Bot Mag 32:151–154
Salmon A, Ainouche ML, Wendel JF (2005) Genetic and epigenetic consequences of recent hybridization and polyploidy in *Spartina* (Poaceae). Mol Ecol 14:1163–1175
Sax K (1922) Sterility in wheat hybrids II chromosome behaviour in partially sterile hybrids. Genetics 7:543–552
Shaked H, Kashkush K, Ozkan H, Feldman M, Levy AA (2001) Sequence elimination and cytosine methylation are rapid and reproducible responses of the genome to wide hybridization and allopolyploidy in Wheat. Plant Cell 13:1749–1759
Soltis DE, Soltis PS (1993) Molecular data and the dynamic nature of polyploidy. Crit Rev Plant Sci 12:243–273
Soltis PS, Soltis DE (2009) The role of hybridization in plant speciation. Annu Rev Plant Biol 60:561–588
Soltis DE, Soltis PS, Pires JC, Kovarik A, Tate JA, Mavrodiev E (2004) Recent and recurrent polyploidy in *Tragopogon* (Asteraceae): cytogenetic, genomic and genetic comparisons. Biol J Linn Soc 82:485–501
Soltis DE, Soltis PS, Schemske DW, Hancock J, Thompson J et al (2007) Autopolyploidy in angiosperms: have we grossly underestimated the number of species? Taxon 56:13–30
Song K, PL L, Tang K, Osborn TC (1995) Rapid genome change in synthetic polyploids of *Brassica* and its implications for polyploid evolution. Proc Natl Acad Sci U S A 92:7719–7723
Stace CE (1989) Plant taxonomy and biosystematics. E. Arnold, Cambridge

Stace CA (2000) Cytology and cytogenetics as a fundamental taxonomic resource for the 20th and 21st centuries. Taxon 49:451–477

Stearns SC (1992) The evolution of life histories. Oxford University Press, Oxford

Stebbins GL (1940) The significance of polyploidy in plant evolution. Am Nat 74:54–66

Stebbins GL (1947) Types of polyploids: their classification and significance. Adv Genet 1:403–429

Stebbins GL (1950) Variation and evolution in plants. Columbia University Press, New York

Stebbins GL (1971) Chromosomal evolution in higher plants. Edward Arnold, London

Strother JL, Brown LE (1988) Dysploidy in hymenoxys texana (compositae). Am J Bot 75:1097–1098

Tackholm G (1922) Zytologische Studien uber die Gattung Rosa. Acta Horti Bergiani Bd 7(3):97–381

Tahara M (1915) Cytological studies on *Chrysanthemum*. Bot Mag Tokyo 29:48–50

Thomas PT (1940a) Reproductive versatility in *Rubus*. 11. The chromosomes and development. J Genet 40:119–120

Thomas PT (1940b) The origin of new forms in *Rubus*. 111. The chromosome constitution of *R. loganobaccus* BAILEY, its parents and derivatives. Ibid 40:141–156

Town CD, Cheung F, Maiti R, Crabtree J, Haas BJ, Wortman JR, Hine EE, Althoff R, Arbogast TS, Tallon LJ, Vigouroux M, Trick M, Bancroft I (2006) Comparative genomics of *Brassica oleracea* and *Arabidopsis thaliana* Reveal gene loss, fragmentation, and dispersal after polyploidy. Plant Cell 18:1348–1359

Tuskan GA, Difazio S, Jansson S, Bohlmann J, Grigoriev I, Hellsten U, Putnam N, Ralph S, Rombauts S, Salamov A, Schein J, Sterck L, Aerts A, Bhalerao RR, Bhalerao RP, Blaudez D, Boerjan W, Brun A, Brunner A, Busov V, Campbell M, Carlson J, Chalot M, Chapman J, Chen GL, Cooper D, Coutinho PM, Couturier J, Covert S, Cronk Q, Cunningham R, Davis J, Degroeve S, Déjardin A, dePamphilis C, Detter J, Dirks B, Dubchak I, Duplessis S, Ehlting J, Ellis B, Gendler K, Goodstein D, Gribskov M, Grimwood J, Groover A, Gunter L, Hamberger B, Heinze B, Helariutta Y, Henrissat B, Holligan D, Holt R, Huang W, Islam-Faridi N, Jones S, Jones-Rhoades M, Jorgensen R, Joshi C, Kangasjärvi J, Karlsson J, Kelleher C, Kirkpatrick R, Kirst M, Kohler A, Kalluri U, Larimer F, Leebens-Mack J, Leplé JC, Locascio P, Lou Y, Lucas S, Martin F, Montanini B, Napoli C, Nelson DR, Nelson C, Nieminen K, Nilsson O, Pereda V, Peter G, Philippe R, Pilate G, Poliakov A, Razumovskaya J, Richardson P, Rinaldi C, Ritland K, Rouzé P, Ryaboy D, Schmutz J, Schrader J, Segerman B, Shin H, Siddiqui A, Sterky F, Terry A, Tsai CJ, Uberbacher E, Unneberg P, Vahala J, Wall K, Wessler S, Yang G, Yin T, Douglas C, Marra M, Sandberg G, Van de Peer Y, Rokhsar D (2006) The genome of black cotton wood *Populus trichocarpa* (Torr.& Gray). Science 313:1596–1604

Udall JA, Wendel JF (2006) Polyploidy and crop improvement. Crop Sci 46:3–14

Vamosi JC, Dickinson TA (2006) Polyploidy and diversification: a phylogenetic investigation in Rosaceae. Int J Plant Sci 167(2):349–358

Wagner WH Jr (1970) Biosystematics and evolutionary noise. Taxon 19:146

Wang J, Tian L, Madlung A, Lee HS, Chen M, Lee JJ, Watson B, Kagochi T, Comai L, Chen ZJ (2004) Stochastic and epigenetic changes of gene expression in *Arabidopsis* polyploids. Genetics 167:1961–1973

Wendel JF, Albert VA (1992) Phylogenetics of the cotton genus (Gossypium): character-state weighted parsimony analysis of chloroplast-DNA restriction site data and its systematic and biogeographic implications. Syst Bot 17(1):115–143. https://doi.org/10.2307/2419069

Winge O (1917) The Chromosomes: their number and general importance. CRTrav LoborCarlsberg 13:131

Winkler H (1916) Uber die experimentelle Erzeugung von Pflanzen mit abweichenden Chromosomenzahlen. Zeitschi f. Bot., Bd. 8

Yarnell SH (1936) Chromosome behaviour in blackberry-raspberry hybrids. J Agric Res 52:385–396

Zhang H, Zhu B, Qi B, Gou X, Dong Y, Xu C, Zhang B, Huang W, Liu C, Wang X, Yang C, Zhou H, Kashkush K, Feldman M, Wendel JF, Bao L (2014) Evolution of the BBAA component of bread wheat during its history at the allohexaploid level. Plant Cell 26:2761–2776

Origin of Polyploidy

2

2.1 Introduction

In terms of time, polyploidy is believed to have accompanied the evolution of plants since the last 350 million years (Vision et al. 2000; Bowers et al. 2003). Since polyploids are known as successful invaders of newly opened habitats, it is hypothesized that the major decline and extinction of gymnosperms during the late Cretaceous period (some 65 million years ago) left room for the newly formed polyploids of woody angiosperms (Stebbins 1950). As of now, it has been reported in almost all groups of plants, being one of the best known of their evolutionary processes.

The methods of origin of polyploidy in plants per se are various. These include somatic doubling during mitosis, gametic non-reduction and polyspermy. Somatic doubling occurs in a zygote or young embryo to produce a completely polyploid individual or, locally in some apical meristems, to give polyploid chimaeras (Ahloowalia and Garber 1961; Hiesey et al. 1971). Polyploids are formed when mitotic or meiotic irregularities cause the formation of gametes that have more than one set of chromosomes. Diploid gametes, which are formed infrequently, fuse with haploid gametes and produce triploid zygotes. These are unstable and can be either sterile or form polyploidy gametes (Fig. 2.1). Polyploidy has also been known to arise from polyembryonic seeds (Webber 1940). Gametic non-reduction during micro- and mega-sporogenesis is the second major mode of polyploid formation. Polyspermy, the fertilization of an egg by more than one sperm nucleus, is observed in many plants, but its contribution as a mechanism for polyploid formation is rather rare, except in some orchids (Ramsey and Schemske 1998).

Triploid bridge is the route of polyploid formation, which involves non-reduction in one of the germ lines (pollens or the eggs). Tetraploid individual is formed by a two-step process, in which the unreduced $2n$ gamete fuses with a reduced 1n gamete to give a $3n$ zygote, followed by the subsequent fusion of a $3n$ gamete with a normal 1n gamete in the next generation to give rise to tetraploid individuals, for example, in artificial *Galeopsis tetrahit* (Grant 1981).

T.-Ul.-H. Dar, R.-Ul. Rehman, *Polyploidy: Recent Trends and Future Perspectives*, https://doi.org/10.1007/978-81-322-3772-3_2

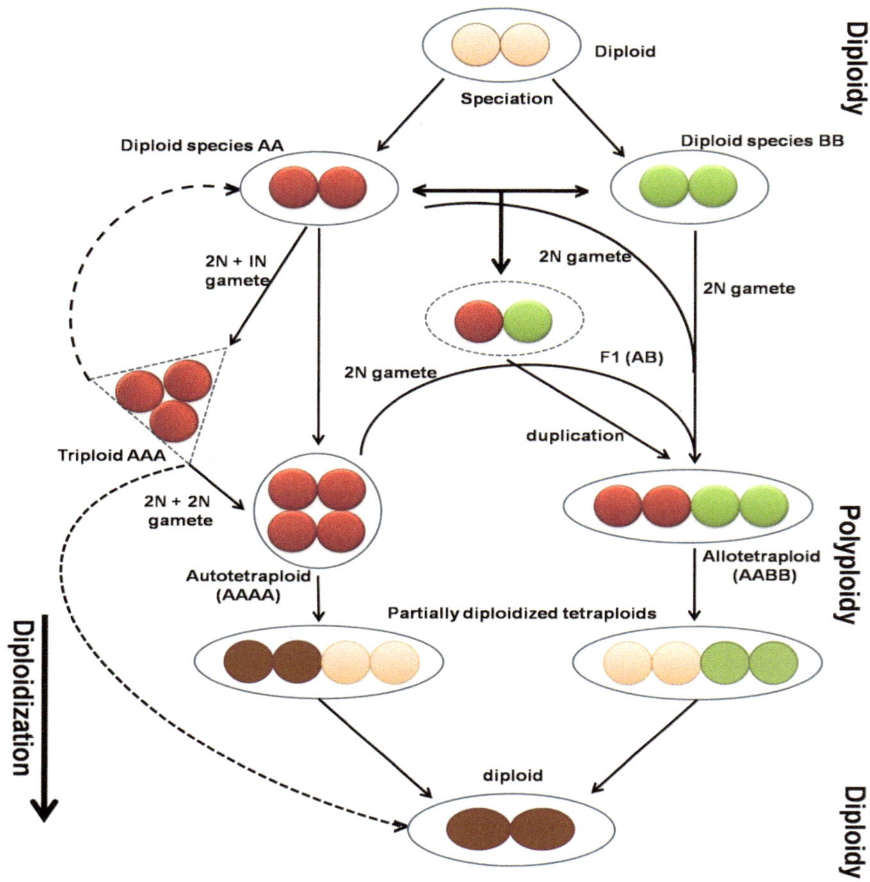

Fig. 2.1 Polyploidization and diploidization
This illustration explains the possible paths that results the sudden transition from diploidy to polyploidy and vice versa. Circles with different colours represent different genomes. Unstable ploidy is represented by dashed lines. A and B represent different genomes and N is gametic number. Diploidization occurs as a result of retention and sub-functionalization of duplicates which in turn increases gene number. Genomes that are shown by ovals represent increased gene number due to diploidization (Modified from Comai 2005)

Polyploids, initially, were considered to have a single origin, resulting in genetic uniformity across all individuals of a species. The application of the recent molecular biology techniques, more especially DNA markers, has demonstrated that a single polyploid species may have multiple origins, leading to more variation in polyploids than would be the case with single origin (Soltis and Soltis 1993a, b, 1999; Leitch and Bennett 1997; Soltis et al. 2004; Doyle et al. 2004; Soltis et al. 2004; Martelotto et al. 2007) (Fig. 2.2a, b).

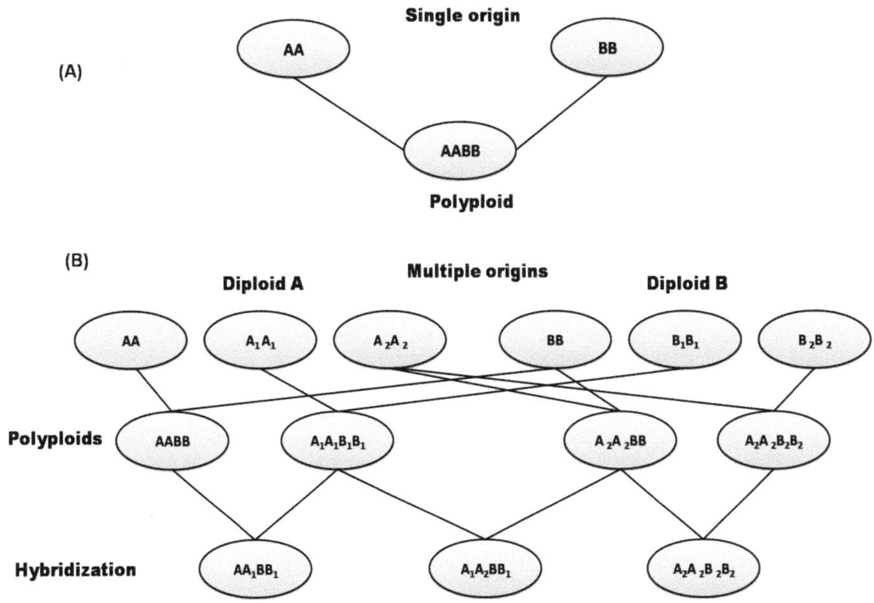

Fig. 2.2 Polyploid formation. (**a**) Traditional view for the formation of genetically uniform. polyploid species. (**b**) Recent view for recurrent formation of polyploid species from different genotypes. Variability in these polyploids occurs due to hybridization, independent assortment and recombination (Modified from Soltis et al. 2004)

2.2 Pathways of Polyploid Formation

Polyploidy is formed by different routes (Ramsey and Schemske 1998). The major routes are given below:

2.2.1 Autotetraploid, Triploid Bridge

Triploids are formed within a diploid population by the union of reduced (*n*) and unreduced (2*n*) gametes. Backcrossing of triploid to diploids or self-fertilization of triploids produces tetraploids (Fig. 2.3). Only a small number of tetraploid progeny were obtained from triploid apple varieties that had originated themselves as spontaneous polyploids (14). It has been observed that gametes produced by spontaneous triploids are not functional due to aneuploidy and unbalanced chromosome numbers (Belling and Blakeslee 1924a, b; Dermen 1931; Dujardin and Hanna 1988; King 1933; Satina and Blakeslee 1937).

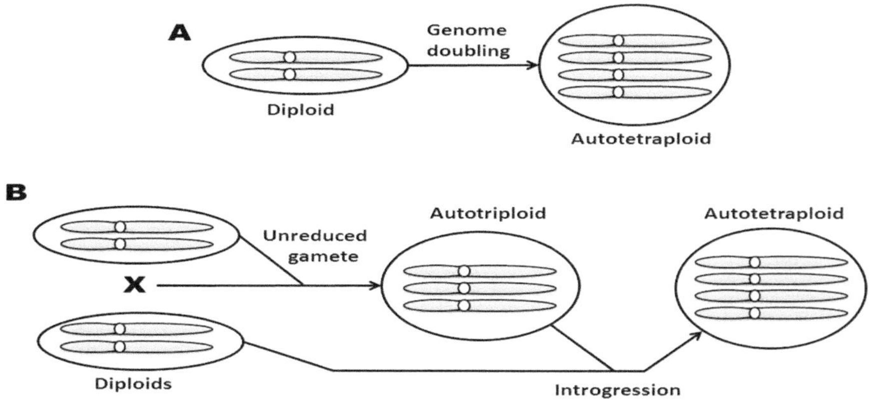

Fig. 2.3 (**a**) Autopolyploidy resulting from spontaneous doubling of the same genome. (**b**) Autotetraploid formation due to fusion of unreduced 2n gamete with a normally reduced gamete within a single species. Fertilization of viable triploid eggs (if produced) with a normally reduced pollen grain produces viable autotetraploid (Modified from Hegarty and Hiscock 2008)

2.2.2 Autotetraploid, One Step

Tetraploids are formed directly in a diploid population by the union of two unreduced (2*n*) gametes or by somatic doubling. Somatic doubling occurs in a zygote or young embryo to produce a completely polyploid individual or, locally in some apical meristems, to give polyploid chimaeras (Ahloowalia and Garber 1961; Hiesey et al. 1971). *Primula kewensis,* one of the first described allopolyploids, originated from somatic doubling in certain flowering branches of a diploid hybrid (Newton and Pellew 1929). In some non-meristematic plant tissues, somatic polyploidy has been seen (Amato 1952, 1964). For example, in the cortex and pith of *Vicia faba,* tetraploid and octoploid cells have been observed (Ramsey and Schemske 1998). The main cause of somatic doubling is mitotic non-disjunction (Grant 1981); this may occur in purely vegetative tissues, or in branches that may produce flowers, or in early embryos. Heat shock experiments were performed in order to explain the phenomenon of somatic doubling in the zygotes. In this experiment young embryos were briefly exposed to high temperatures that lead to somatic doubling (Randolph 1932; Dorsey 1936).

Gametic non-reduction during micro- and mega-sporogenesis is the second major mode of polyploid formation. This non-reduction could be due to meiotic non-disjunction (failure of the chromosome to separate and subsequent reduction in chromosome number), formation of gametes by mitosis instead of meiosis or failure of cell wall formation. The union of reduced and unreduced gametes can generate polyploid embryos (Fig. 2.4).

2.2.3 Allotetraploid, Triploid Bridge

In this case, allotetraploids are formed by hybrid triploids. These hybrid triploids are formed by interspecific crosses of diploids in F1 or F2 generation. Self-fertilization

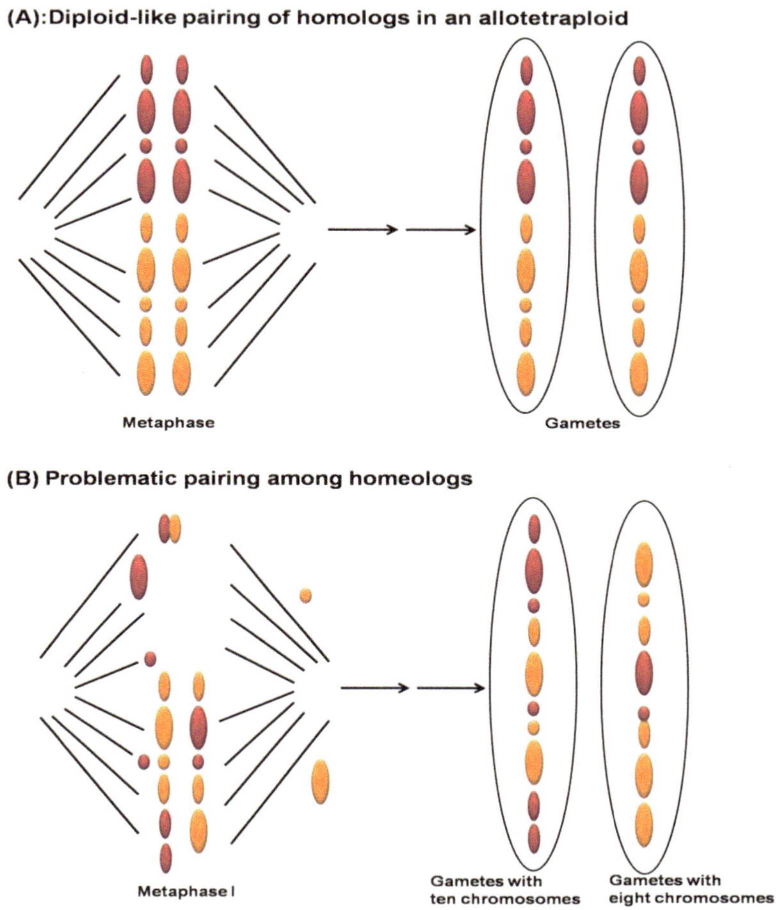

Fig. 2.4 (**a** and **b**). Chromosome pairing and segregation in natural and newly formed allotetraploids
(**a**) Diploid like pairing of homologs in natural allotetraploid at metaphase I of meiosis. Chromosome complements from two different progenitors are indicated with two different colours. Homologs chromosomes pair normally showing permanent heterozygosity
(**b**) Problematic chromosome pairing in newly formed allotetraploids. In newly formed allotetraploid chromosome rearrangements can occur in the progenitors in the time since their divergence from a common ancestor and can lead to different chromosome numbers. Chromosome of one progenitor can show similarity to more than one chromosome from other progenitor and can pair with multiple chromosomes. As a result chromosomes are unequally distributed (Modified from Pikaard 2001)

or backcrossing of hybrid triploid to diploids produces allotetraploids. Müntzing (1932) generated a highly sterile diploid F1-hybrid by crossing *Galeopsis pubescens* and *Galeopsis speciosa*. Out of 200 progenies, only one was found to be triploid. Backcross of this one F2 plant with *G. pubescens* formed a single viable tetraploid seed. Allotriploids have been commonly observed in the F2 generation produced by backcrossing or selfing interspecific F1 hybrids (Buxton and Newton 1928; Clausen et al. 1945; Levan 1941; Müntzing 1930; Poole 1931) and in the F1 generation by the union of reduced and unreduced gametes from the parent genotypes (Collins 1933; Hahn 1990; Skalińska 1945) (Fig. 2.5).

2.2.4 Allotetraploid, One Step

Allotetraploids are directly formed from diploids in the F1 or F2 generation of interspecific crosses. For example, 90% of the F2 progenies of *Digitalis ambigua* and *Digitalis purpurea* were tetraploid (Buxton and Newton 1928). Half of the F2 progenies of *Allium cepa* and *Allium fistolium* were tetraploid or hypotetraploid (Levan 1941), and 2% of the F1 progenies of *Manihot pruinosa glaziovii* were tetraploid (Hahn et al. 1990). There are many other examples of tetraploids being produced in one step by F1 interspecific hybrids (Abdel-Hameed and Snow 1972; Buxton and Newton 1928; Clausen et al. 1940; Grant 1952; Hiesey et al. 1971; Karpechenko 1927; Newton and Pellew 1929; Poole 1931) or in the F1 generation of an interspecific cross (Hahn et al. 1990; Jones and Clarke 1942; Levan 1937). Other important pathways involve the evolution of ploidy levels above tetraploidy.

2.2.5 Higher Ploidy, One Step

The union of reduced and unreduced gametes generates a new cytotype of higher ploidy in a polyploid population. For example, from the progeny of open pollinated autotetraploids, *Beta vulgaris*, 2% hexaploid cytotypes were recovered apparently from the union of reduced ($2x$) and unreduced ($4x$) gametes (Hornsey 1973). Similarly, 1% of the progenies of tetraploid alfalfa were found to be hexaploid (Bingham 1968). There is circumstantial evidence of autohexaploid formation in tetraploid populations in several other systems (Clausen et al. 1940; Dewey and Asay 1975). New odd-ploidy cytotypes could also be produced by this mechanism. For example, it has been suggested that unreduced gamete production in hexaploid *Andropogon gerardii* generated a $9x$ cytotype, which is now widely distributed (Norrmann et al. 1997).

2.2.6 Allopolyploidy, via Hybridization of Autopolyploids

Among distinct autopolyploids hybridization directly produces allopolyploids. For example, crosses between autotetraploid *Lycopersicon esculentum* and autotetraploid *Lycopersicon pimpinellifolium* produced a fertile allotetraploid, *Lycopersicon,* that was identical to the allotetraploid made by doubling the diploid F1 hybrid (Lindstrom and Humphrey 1933). An allotetraploid *Tradescantia* was produced by crossing autotetraploid forms of *T. canaliculata* and *T. subaspera* (Anderson 1936). It has been repeatedly found that the post-zygotic barriers that isolate diploid taxa break down in autopolyploids, so that interspecific hybrids are formed easily

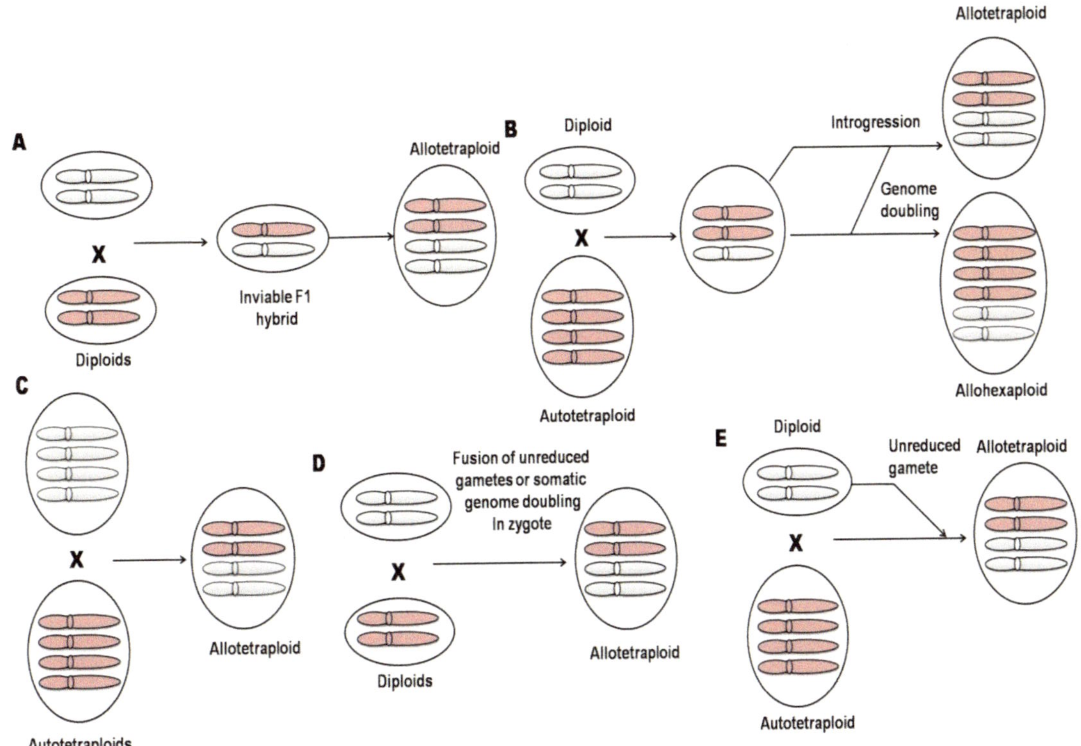

Fig. 2.5 (**a**–**e**) Allopolyploidy
(**a**) Crossing between two species yielding an inviable F1 hybrid due to irregular meiotic pairing. This is followed by genome doubling to produce fertile allotetraploid
(**b**) Crossing between two species (diploid and autotetraploid) produces sterile F1 triploid. This is followed by genome doubling to produce viable allohexaploid. Sometimes triploids may backcross to the diploid parent produces allotetraploid offspring
(**c**) Crossing between two autotetraploids resulting in an allotetraploid
(**d**) An allotetraploid is also produced from fusion of two unreduced gametes or somatic genome doubling in zygotes
(**e**) Crossing between two species involves fusion of an unreduced gamete from a diploid and reduced from a autotetraploid (Modified from Hegarty and Hiscock 2008)

(Clausen et al. 1940; Heckard 1960). Not surprisingly, there may be extensive intergradation among polyploids, while diploid taxa remain morphologically distinct (Heckard 1960; Lewis and Szweykowski 1964).

2.2.7 Allopolyploidy, via Hybridization of Different Cytotypes

Hybridization among different cytotypes (which may be of auto- or allopolyploid origin) generates intermediates of odd-ploidy, which subsequently produce new even ploidy cytotypes. For example, a high frequency (81%) of allohexaploids was generated by crossing triploid hybrids of *Nicotiana paniculata* (2*x*) and *Nicotiana rustica* (4*x*) (Lammerts 1931). A similar process has been observed in many other systems (Clausen et al. 1945; Lammerts 1931; Thompson 1942; Wagenaar 1968).

2.3 Artificial Induction of Polyploidy

In plants induction of polyploidy has been used for developing new plant characters and also is considered a hot topic for researchers (Cheng and Korban 2011; Dhooghe et al. 2011). Polyploidy can also be induced artificially. Different methods which are utilized for the induction of polyploidy are given below:

1. By Radiation:

   Radiations of shorter wavelengths, e.g. X-rays, gamma rays and ultraviolent rays, have been used for the induction of polyploidy (Breslavetz 1939; Nettancourt et al. 1971; Särkilahti 1988). Plant parts such as vegetative and flower buds can be exposed to radiations for brief timing. Irradiation induces somatic doubling of chromosomes and also rate of cell division.
2. By Temperature Shocks:

   Low- and high-temperature shocks have also been utilized for the induction of polyploidy in many studies. In Datura cold treatment at zygotic stage was used for the induction of polyploidy (Belling and Blakeslee 1924a, b). High-temperature treatment was used in maize, barley, rice, wheat sweet clover, *Capsicum* and *Populus* for inducing polyploidy (Randolph 1932; Matusima 1935; Dorsey 1936; Atwood 1936; Peto 1938; Kang et al. 2000; Zhang et al. 2002; Wang et al. 2012; Li et al. 2017 and Lu et al. 2013).
3. By Injury:

   Plant cells have a property of totipotency. When the plant part usually meristematic zone is injured, the cells at that point grow rapidly and form a callus. Coumarin, a chemical compound produced by plants, induced the callus growth which also leads to somatic doubling of chromosomes. Vegetative buds developed from the callus are generally polyploids. Tetraploids were produced in tomato, *Solanum* and *Nicotiana* by this method.
4. Use of Unreduced Pollen:

Unreduced diploid pollens with their larger size can be picked by fine glass tubing and can be used for pollination. The seeds produced could be screened for polyploidy. Pratassenja (1939) used this method in peaches. Recently, Yao et al. (2017) induced unreduced pollen by using high temperature in rubber tree clone RRIM 600.

5. Production of Diploid Gametophytes:

   Polyploidy in some mosses and ferns could be obtained by using a diploid tissue from sporophyte to regenerate into a gametophyte. Tetraploid sporophyte will be generated from diploid gametophyte by normal process of sexual reproduction.

6. Production of Polyploidy by Nitrous Oxide:

   Nitrous oxide can also be used for production of artificial polyploids. In this method zygotes and proembryo (24–48 hours after pollination) are treated with nitrous oxide (N_2O With 3–6 atmosphere). Pollinated spikes are taken and placed in flasks containing liquid nutrient medium for 5–24 hours are used. These spikes are kept in growth chamber for 2 weeks, after that seedling are obtained through embryo culture. This method has been used in many plants, *Crepis capillaris* (Ostergren 1954), *Phalaris* (Ostergren 1957), *Melandrium* (Nygren 1955), wheat and barley (Dvorak et al. 1973), *Trifolium* (Taylor et al. 1976), maize (Kato 2002) and hybrid lilies (Akutsu et al. 2007; Kitamura et al. 2009).

   Nitrous oxide gas produces polyploidy by depolymerizing microtubules; as a result chromosomes do not move to opposite poles and remained in the centre of the cell. N_2O treatment for the induction of polyploidy has several advantages over colchicine treatment. First, colchicine is generally applied to the seedlings, seeds or growing plants which consists of a large number of cells. As a result chimaeras of both diploid and tetraploid cells are formed, and for a true polyploid, another generation is to be obtained. In comparison N_2O treatment can be applied to pollinated flowers in which first mitosis is just taking place in the fertilized egg cell. Secondary, N_2O treatment produces polyploidy more efficiently than colchicine.

7. Chemical Treatment:

   Various chemicals are used for the induction of polyploidy in plants, e.g. colchicine, oryzalin, trifluralin and amiprophos-methyl (Blakeslee and Avery 1937; Bouvier et al. 1994; Van Tuyl et al. 1992; Dewitte et al. 2010; Podwyszynska 2011). Among them colchicine is the most commonly used one (Rogalska et al. 2007; Dhooghe et al. 2011). Colchicine ($C_{22}H_{25}NO_6$) is an alkaloid extracted from the seeds and corms of *Colchicum autumnale.* It is highly soluble in water. Blakeslee (1937) and Nebel (1937) discovered successful use of colchicine in inducing polyploidy. Colchicine binds to tublin and inhibits polymerization of microtubules; as a result spindle fibres are not formed (Dustin 1984; Wilson et al. 1985; Medrano et al. 1989).

Colchicine is usually used as an aqueous solution in various concentrations. Sometimes different carriers (glycerine, lanolin and nutrient agar) may be used in order to increase the effectiveness of treatment. The aqueous solution of colchicine is applied to meristematic tissues. The concentration and time duration

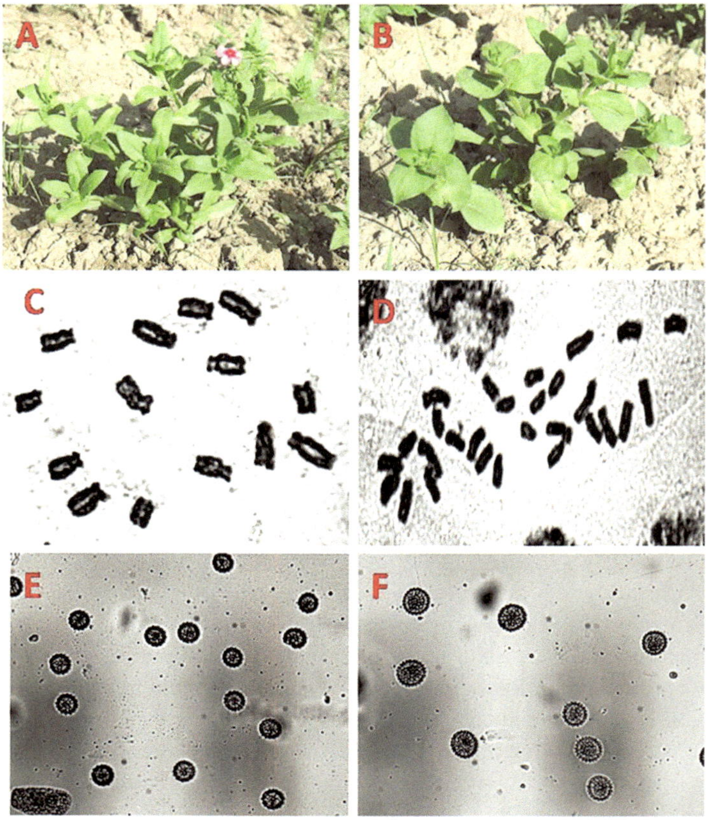

Fig. 2.6 Comparison between diploid (left) and colchitetraploid (right) Phloxdrummondii. Phenotypic variation between diploid (**a**) and colchitetraploid (**b**) plants. Robust plant growth and larger leaf size are seen in autotetraploid. Mitotic root tip spread of diploid (**c**) and colchitetraploid (**d**). Larger pollen size in colchitetraploid as compared to diploid (**e** and **f**)

vary from material to material. Colchicine has been used by many investigators for chromosome doubling in many crop plants (Blakeslee and Avery 1937; Dermen 1938, 1947, 1954; Clausen 1941; Gao et al. 1996; Gauthier et al. 1998; Gao et al. 2002; Viehmannová et al. 2009; Mishra et al. 2010; Zhang et al. 2010; Podwyszynska et al. 2011; Trojak-Goluch and Skomra 2013; Dar et al. 2017). The colchitetraploids obtained are analysed by using different parameters, e.g. leaf texture, leaf size, stomatal size, pollen grains and meiotic as well as mitotic analysis. Initially colchitetraploids show slow growth but later they achieve robust growth. Leaf size in colchitetraploids is larger as compared to diploids. Stomatal and pollen size are also larger in colchitetraploids as compared to diploids (Dar et al. 2017) (Fig. 2.6).

Methods of Colchicine Treatment

1. Seed and Protocorm Treatment:

   The seeds are first soaked in water for 10–24 h and then transferred to colchicine solution. Concentration of aqueous solution ranged from 0.05 to 1.0 per cent with time duration of 1–10 days. Seeds may also be sown into colchicine-incorporated nutrient agar. In case of protocorm, seeds are first sown and germinated in nutrient agar. When green protocorms are formed, they are transferred to a flask and approximately 0.01–1.5% of sterilized colchicine is added under aseptic conditions (Nakasone and Kamemoto 1961).
2. Seedling Treatment:

   This method is more effective than seed treatment. The seeds are first allowed to germinate on moist blotting paper or cotton, and then germinated seedlings are soaked in 0.01–2.0% aqueous colchicine solution for 2–24 h under two methods. First, the method involves the immersion of seedlings for 3–5 h in vials containing colchicine solution. The roots of seedlings should be protected from colchicine and should be kept moist by wrapping wet cotton strips. The second method is known as infiltration method (Braak and Zeilinga 1957). In this method immersed seedling in vials containing aqueous colchicine solution is kept in a desiccator in which a vacuum is created by water vacuum pump. The plants are remained in the vacuum for 10 min. The air from plants is evacuated and the colchicine solution penetrates the tissues more efficiently and rapidly. After treatment, the seedlings are removed from vials and rinsed in tap water.
3. Cotton Swab Method:

   This method is more effective due to low mortality rate. Sterilized cotton balls of suitable size are placed on the emerging apical tips between two cotyledonary leaves of 3–4-day-old seedlings. Then aqueous solution of colchicine is applied drop by drop on the cotton ball with the help of Pasteur pipette. Colchicine treatment is applied 4–8 h continuously for 2–3 days (Dar et al. 2017).
4. Treatment of Inflorescence of Mature Plants:

   Flower spikes 2.5–20 cm long are selected and wrapped with absorbent cotton. The aqueous solution of colchicine (0.1–2.05%) is applied to the cotton. The inflorescence is covered with polythene bags in order to prevent drying. Duration of the treatment is 8 h to 5 days. After treatment cotton is removed and spikes are washed with tap water and allowed to grow.
5. Treatment of Tip Cuttings of Mature Plants:

   Tips of approximately 8–10 in. long are cut and immersed at basal ends in vials containing aqueous solution of colchicine. The duration of the treatment varies from 1 to 20 days and the concentration used is 0.1–2.0%. After treatment cuttings are sown in the field for further development.

References

Abdel-Hameed F, Snow R (1972) The origin of the allotetraploid *Clarkia gracilis*. Evolution 26:74–83

Ahloowalia BS, Garber FD (1961) The genus *Collinsia*. XIII. Cytogenetic studies of interspecific hybrids involving species with pediceled flowers. Bot Gaz 122:219–228

Akutsu M, Kitamura S, Toda R, Miyajima I, Okazaki K (2007) Production of 2*n* pollen of Asiatic hybrid lilies by nitrous oxide treatment. Euphytica 155:143–152

Amato DF (1952) Polyploidy in the differentiation and function of tissues and cells in plants. Caryologia 4:311–358

Amato DF (1964) Endopolyploidy as a factor in plant tissue development. Caryologia 17:41–52

Anderson E (1936) A morphological comparison of triploid and tetraploid interspecific hybrids in *Tradescantia*. Genetics 21:61–65

Atwood SS (1936) Tetraploid and aneuploidy *Melilotus alba* resulting from heat treatment. Am J Bot 23:674–677

Belling J, Blakeslee AF (1924a) The configurations and sizes of chromosomes in the trivalents of 25-chromosome Daturas. Proc Natl Acad Sci U S A 10:116–120

Belling J, Blakeslee AF (1924b) The distribution of chromosomes in tetraploid Daturas. Am Nat 58:60–70

Bingham ET (1968) Aneuploids in seedling populations of tetraploid alfalfa, *Medicago sativa* L. Crop Sci 8:571–574

Blakeslee AF (1937) dédoublement du nombre des chromosomes chez les plantes par traitment chimique. Compt Rend Acad Sci Paris 205:476–479

Blakeslee AF, Avery AG (1937) Method of inducing doubling of chromosomes in plants by treatment with colchicine. J Hered 28:393–411

Bouvier L, Fillon FR, Lespinasse Y (1994) Oryzalin as an efficient agent for chromosome doubling of haploid apple shoots *in vitro*. Plant Breed 113:343–346

Bowers JE, Chapman BA, Rong J, Paterson AH (2003) Unravelling angiosperm genome evolution by phylogenetic analysis of chromosomal duplication events. Nature 422:433–438

Braak JP, Zeilinga AE (1957) Production of a colchicine-induced tetraploid *Asparagus*. Euphytica 6(20):1–212

Breslavetz L (1939) Polyploids in rye induced by X-rays. Akad Nauk (Doklady) Moscow SSSR 22:354–357

Buxton BH, Newton WCF (1928) Hybrids of *Digitalis ambigua* and *Digitalis purpurea*, their fertility and cytology. J Genet 19:1269–1279

Cheng ZM, Korban SS (2011) In vitro ploidy manipulation in the genomics era. Plant Cell Tissue Organ Cult 104:281–282

Clausen RE (1941) Polyploidy in *Nicotiana*. Am Nat 75:291–306

Clausen J, Keck DD, Hiesey WM (1940) Experimental studies on the nature of species.I. Effect of varied environments on western North American plants. Carnegie Institute of Washington, Publ. No. 520, pp 1–452

Clausen J, Keck DD, Hiesey WM (1945) Experimental studies on the nature of species. II. Plant evolution through amphiploidy and autoploidy, with examples from the Madiinae. Carnegie Institute of Washington, Publ. No. 564

Collins JL (1933) Morphological and cytological characteristics of triploid pineapples. Cytologia 26:248–256

Comai L (2005) The advantages and disadvantages of being polyploidy. Nat Rev Genet 6:836–846

Dar TH, Raina SN, Goel S (2017) Cytogenetic and molecular evidences revealing genomic changes after autopolyploidization: a case study of synthetic autotetraploid *Phlox drummondii* hook. Physiol Mol Biol Plants. https://doi.org/10.1007/s12298-017-0445-8

Dermen H (1931) Polyploidy in petunia. Am J Bot 18:250–261

Dermen H (1938) A cytological analysis of polyploidy induced by colchicine and by extremes of temperature. J Heredity 29:211–229

Dermen H (1947) Inducing polyploidy in peach varieties. J Hered 38:77–82
Dermen H (1954) Colchicine in grapes. J Hered 45:159–172
Dewey DR, Asay KH (1975) The crested wheat grasses of Iran. Crop Sci 15:844–849
Dewitte A, Eeckhaut T, Van Huylenbroeck J, Van Bockstaele E (2010) Induction of 2n pollen formation in Begonia by trifluralin and N2O treatments. Euphytica 171:283–293
Dhooghe E, Van Laere K, Eeckhaut T, Leus L, Van Huylenbroeck J (2011) Mitotic chromosome doubling of plant tissues *in vitro*. Plant Cell Tissue Organ Cult 104:359–373
Dorsey E (1936) Induced polyploidy in wheat and rye. Chromosome doubling in *Triticum*, *Secale* and *Triticum-Secale* hybrids produced by temperature changes. J Hered 27:155–160
Doyle JJ, Doyle JL, Rauscher J, Brown AHD (2004) Evolution of gene families and gene conversion in the *Glycine* polyploid complex. Biol J Linn Soc 82(4):583–597
Dujardin M, Hanna WW (1988) Cytology and breeding behavior of a partially fertile triploid pearl millet. J Hered 79:216–218
Dustin P (1984) Microtubules. Springer, New York
Dvorak J, Harvey BL, Coulman BE (1973) The use of nitrous oxide for producing euploids and aneuploids in wheat and barley. Can J Genet Cytol 15:205–214
Gao SL, Zhu DN, Cai ZH, DR X (1996) Autotetraploid plants from colchicine-treated bud culture of *Salvia miltiorrhiza*Bge. Plant Cell Tissue Organ Cult 47:73–77
Gao SL, Chen BJ, Zhu DN (2002) In vitro production and identification of autotetraploids of *Scutellaria baicalensis*. Plant Cell Tissue Organ Cult 70:289–293
Gauthier P, Lumaret R, Bedecarrats A (1998) Genetic variation and gene flow in alpine diploid and tetraploid populations of Lotus. I. Insights from morphological and allozyme markers. Heredity 80:683–693
Grant V (1952) Cytogenetics of the hybrid *Gilia millefoliata* × *Achilleaefolia*.I.Variations in meiosis and polyploidy rate as affected by nutritional and genetic conditions. Chromosoma 5:372–390
Grant V (1981) Plant speciation, 2nd edn. Columbia University Press, New York
Hahn SK, Bai KV, Asiedu R (1990) Tetraploids, triploids, and 2*n* pollen from diploid interspecific crosses with cassava. Theor Appl Genet 79:433–439
Heckard LR (1960) Taxonomic studies in the *Phacelia magellanica* polyploid complex. Univ Calif Publ Bot 32:1–126
Hegarty MJ, Hiscock SJ (2008) Genomic clues to the evolutionary success of polyploid plants. Curr Biol 18:R435–R444
Hiesey WM, Nobs MA, Björkman O (1971) Experimental studies on the nature of species. V. Biosystematics, genetics, and physiological ecology of the *Erythranthe* section of *Mimulus*. Carnegie Institute of Washington Publ 628, pp 1–213
Hornsey KG (1973) The occurrence of hexaploid plants among autotetraploids populations of sugar beet (*Beta vulgaris* L.), and the production of tetraploid progeny using a diploid pollinator. Caryologia 26:225–228
Jones HA, Clarke AF (1942) A natural amphi diploid from an onion species hybrid. J Hered 33:25–32
Kang XY, Zhu ZT, Zhang ZY (2000) Suitable period of high temperature treatment for 2n pollen of *Populus tomentosa* × *P. bolleana*. J Beijing For Univ 22:1–4
Karpechenko GD (1927) The production of polyploid gametes in hybrids. Hereditas 9:349–368
Kato A (2002) Chromosome doubling of haploid maize seedling using nitrous oxide gas at the flower primordial stage. Plant Breed 121:370–377
King E (1933) Chromosome behavior in a triploid *Tradescantia*. J Hered 24:253–256
Kitamura S, Akutsu M, Okazaki K (2009) Mechanism of action of nitrous oxide gas applied as a polyploidizing agent during meiosis in lilies. Sex Plant Reprod 22:9–14
Lammerts WE (1931) Interspecific hybridization in *Nicotiana*. XII. The amphidiploid *rustica paniculata* hybrid: its origin and cytogenetic behavior. Genetics 16:191–211
Leitch IL, Bennett MD (1997) Polyploidy in angiosperms. Trends Plant Sci 2:470–476
Levan A (1937) Cytological studies in the *Allium paniculatum* group. Hereditas 23:317–370

Levan A (1941) The cytology of the species hybrid *Allium cepa fistulosum* and its polyploid derivatives. Hereditas 27:253–272

Lewis H, Szweykowski J (1964) The genus *Gayophytum* (Onagraceae). Brittonia 16:343–391

Li Y, Tian M, Zhang P (2017) Embryo sac chromosome doubling in Populus alba × P. Glandulosa induced by high temperature exposure to produce triploids. Breed Sci 267(3):233–238

Lindstrom EW, Humphrey LM (1933) Comparative cyto-genetic studies of tetraploid tomatoes from different origins. Genetics 18:193–209

Lu M, Zhang P, Kang X (2013) Induction of 2n female gametes in *Populus adenopoda* Maxim by high temperature exposure during female gametophyte development. Breed Sci 63(1):96–103

Martelotto LG, Ortiz JPA, Stein J, Espinoza F, Quarin CL, Pessino SC (2007) Genome rearrangements derived from auto polyploidization in *Paspalum* sp. Plant Sci 172:970–977

Matusima K (1935) Chromosome doubling in the rice plants treated with high temperature. Jpn J Genet 11:235–237

Medrano FJ, Andreu JM, Gorbunoff MJ, Timasheff SN (1989) Role of colchicine rings B and C in the binding process to tubulin. Biochemistry 28:5589–5599

Mishra BK, Pathak S, Sharma A, Trivedi PK, Shukla S (2010) Modulated gene expression in newly synthesized auto-tetraploid of *Papaver somniferum* L. S Afr J Bot 76:447–452

Müntzing A (1930) Outlines to a genetic monograph of the genus *Galeopsis* with special reference to the nature and inheritance of partial sterility. Hereditas 13:185–341

Müntzing A (1932) Cyto-genetic investigations on synthetic *Galeopsis tetrahit*. Hereditas 16:105–154

Nakasone HY, Kamemoto H (1961) Artificial induction of polyploidy in orchids by the use of colchicine. Technical Bulletin No 42, Hawaii Agricultural, Experiment Station, University of Hawaii

Nebel BR (1937) Mechanism of polyploidy through colchicine. Nature 140:1101

Nettancourt DD, Dijkhuis P, Gastel AV, Broertjes C (1971) The combined use of leaf irradiation and of the adventitious bud technique for inducing and detecting polyploidy, marker mutations and self-compatibility in clonal populations of *Nicotiana alata* Link and *Otto*. Euphytica 20:508–520

Newton WCF, Pellew C (1929) *Primula kewensis* and its derivatives. J Genet 20:405–467

Norrmann G, Quarín C, Keeler K (1997) Evolutionary implications of meiotic chromosome behavior, reproductive biology, and hybridization in 6x and 9x cytotypes of *Andropogon gerardii* (Poaceae). Am J Bot 84:201–208

Nygren A (1955) Polyploids in *Melandrium* produced by nitrous oxide. Hereditas 41:287–290

Ostergren G (1954) Polyploids and aneuploids of *Crepis capillaris* produced by treatment with nitrous oxide. Genetica 27:54–64

Ostergren G (1957) Production of polyploids and aneuploids of *Phalaris* by means of nitrous oxide. Hereditas 43:512–516

Peto FH (1938) Hybridization of *Triticum* and *Agropyron* V. Doubling the chromosome number in *T. vulgare* and F1 of *T. vulgare* X *A. glaucum* by temperature treatments. Can J Res 16:516–529

Pikaard CS (2001) Genomic changes and gene silencing in polyploids. Trends Genet 17(12):675–677

Podwyszynska M, Gabryszewska E, Jasinski A, Sochacki D (2011) Histogenic identification by cytological analysis of colchicine-induced polyploids of Hemerocallis. Acta Hortic (886):245–250

Poole CF (1931) The interspecific hybrid, *Crepis rubra* ×*C. foetida*, and some of its derivatives. I. Univ Calif Publ Agric Sci 6:169–200

Pratassenja GD (1939) Production of polyploid plants: haploids and triploids in *Prunus persica*. Akademiia Nauk (Doklady), N.S. SSSR 22:348–351

Ramsey J, Schemske DW (1998) Pathway, mechanism and rates of polyploidy formation in flowering plants. Ann Rev Eco Syst 29:467–501

Randolph LF (1932) Some effects of high temperature on polyploidy and other variations in maize. Proc Natl Acad Sci U S A 18:222–229

Rogalska SM, Achrem M, Kalinka A (2007) Mechanisms of genome changes and gene expression in plant hybrid polyploids. Kosmos 56:421–433

Särkilahti E (1988) Micropropagation of a mature colchicine-polyploid and irradiation-mutant of *Betula pendula* Roth. Tree Physiol 14:173–179

Satina S, Blakeslee AF (1937) Chromosome behavior in triploids of *Datura stramonium*.I. The male gametophyte. Am J Bot 24:518–527

Skalińska M (1945) Cytogenetic studies in triploid hybrids of *Aquilegia*. J Genet 47:87–111

Soltis DE, Soltis PS (1993a) Molecular data facilitate a reevaluation of traditional tenets of polyploidy evolution. Crit Rev Plant Sci 12:243–273

Soltis DE, Soltis PS (1993b) Molecular data and the dynamic nature of polyploidy. Crit Rev Plant Sci 12:243–273

Soltis DE, Soltis PS (1999) Polyploidy: origins of species and genomic evolution. Trends Ecol Evol 9:348–352

Soltis DE, Soltis PS, Pires JC, Kovarik A, Tate J, Mavrodiev E (2004) Recent and recurrent polyploidy in *Tragopogon* (Asteraceae): cytogenetic, genomic and genetic comparisons. Biol J Linn Soc 82:485–501

Stebbins GL (1950) Variation and evolution in plants. Columbia University Press, New York

Taylor NL, Anderson MK, Quesenbery KH, Watson L (1976) Doubling the chromosome number of *Trifolium* species using nitrous oxide. Crop Sci 16:516–518

Thompson RC (1942) An amphidiploids *Lactuca*. J Hered 33:253–264

Trojak-Goluch A, Skomra U (2013) Artificially induced polyploidization in *Humulus lupulus* L. and its effect on morphological and chemical traits. Breed Sci 63:393–399

Van Tuyl JM, Meijer B, Van D¨nMP (1992) The use of oryzalin as an alternative for colchicine *in vitro* chromosome doubling of *Lilium* and *Nerine*. Acta Hort 325:625–630

Viehmannová I, Cusimamani EF, Bechyne M, Vyvadilová M, Greplová M (2009) *In vitro* induction of polyploidy in yacon (*Smallanthus sonchifolius*). Plant Cell Tissue Organ Cult 97:21–25

Vision TJ, Brown DG, Tanksley SD (2000) The origins of genomic duplications in *Arabidopsis*. Science 290:2114–2117

Wagenaar EB (1968) Meiotic restitution and the origin of polyploidy. I. Influence of genotype on polyploid seed set in a *Triticum crassum* × *T. turgidum* hybrid. Can J Genet Cytol 10:836–843

Wang J, Li DL, Kang XY (2012) Induction of unreduced megaspores with high temperature during megasporogenesis in *Populus*. Ann For Sci 15:1–9

Webber JM (1940) Polyembryony. Bot Rev 6:575–598

Wilson L, Miller HP, Jordan MA, Farrell KW (1985) In: De Brabander M, De May J (eds) *Microtubules & microtubule inhibitors*. Elsevier Science Publishers B. V, Amsterdam, pp 71–78

Yao PQ, Li GH, Long QY, He LG, Kang XY (2017) Microsporogenesis and induction of unreduced pollen with high temperatures in rubber tree clone RRIM 600. Forests 8:152

Zhang XZ, Liu GJ, Yan LY, Zhao YB, Chang RF, LP W (2002) Creating triploid germplasm via induced 2n pollen in *Capsicum* L. J Horticult Sci Biotechnol 178:84–88

Zhang QY, Luo FX, Liu L, Guo FC (2010) *In vitro* induction of tetraploids in crape myrtle (*Lagerstroemia indica* L.) Plant Cell Tissue Organ Cult 101:41–47

Occurrence of Polyploidy

3

3.1 Introduction

Compared to animals, polyploidy is successful in plants because their developmental processes are much simpler (Fankhauser 1945; Stebbins 1950). The actual incidence or frequency in various plant lineages has been proved difficult to determine, despite numerous attempts over the past 100 years to estimate the frequency of polyploidy in plants (Table 3.1). The phenomenon is comparatively little studied in the thallophytes.

3.2 Polyploidy in Algae

Multiple series of chromosome numbers have been reported long back in some genera of algae, viz. *Cladophora*, *Chara* and *Lomentaria* (Tischler 1931, 1938). Now it is said that polyploidy is also common among algae, though estimates of its incidence are rare. Very high levels of polyploidy may occur in algae, as chromosome numbers from 8 to over 500 are known; besides, allopolyploidy is regarded as a major factor in the evolution of some algal groups (Nichols 1980). According to Lewis (1980) polyploidy is well known in algae and charophytes, but uncommon in the brown and red algae (Table 3.2).

3.3 Polyploidy in Bryophytes

In bryophytes, mosses have been well-known examples of polyploidy, both natural and artificially induced (Wettstein 1927), but it is uncommon in liverworts (Grant 1981). Whereas many authors accept that a high proportion of about 80% of bryophytes are polyploids, assuming low basic chromosome numbers of 5–7 in mosses (Newton 1984; Kuta and Przywara 1997; Smith 1978; Anderson 1980), others challenge this, suggesting that this proportion is much lower (Voglmayr 2000)

T.-Ul.-H. Dar, R.-Ul. Rehman, *Polyploidy: Recent Trends and Future Perspectives*, https://doi.org/10.1007/978-81-322-3772-3_3

Table 3.1 Prevalence of polyploidy in various plant groups

Plant group	Frequency	References
Angiosperms	30–35%	Stebbins (1938, 1950)
	47%	Grant (1963, 1981)
	70%	Masterson (1994)
	2–4%	Otto and Whitton (2000)
	35%	Wood et al. (2009)
Algae	May be very high	Nichlos 1980
Bryophytes	80%	Newton (1984) and Kuta Przywara (1997)
Fungi	Rare or lacking	Stebbins (1950) and Lewis (1980)
Gymnosperms	<5%	Khoshoo 1959
Pteridophytes	95%	Grant (1981) and Otto and Whitton (2000)

Table 3.2 Prevalence of polyploidy in algae

Algal group	Example	Chromosome numbers	References
Chlorophycophyta	*Chlamydomonas eugametos*	16	Buffaloe (1958)
Volvocales	*Astrephomene guberniculifera*	4, 6, 7,8	Cave and Pocock (1956)
	Eudorina elegans	24	Goldstein (1964)
	Eudorina illinoisensis	28	
Zygnematales	*Cosmarium turpinii*	MV	Starr (1958)
	Cosmarium botrytis	MV	Brandham and Godward (1964)
	Staurastrum dilatum	103	Brandham (1965)
	Pleurotaenium mamillatum	53	Ling and Tyler (1976)
	Spirogyra sp.	5–13 (variable)	Godward (1966, 1950)
Cladophorales	*Cladophora glomerata*	Based on 12: 24, 36 96,72,	Sinha (1958)
	Ceramium fimbriatum, *Ceramium cruciatum*	48 42, 84	Rao et al. (1977)
Phaeophycophyta	*Ectocarpus siliculosus*	20, 25	Muller (1976)
Pyrrophycophyta	*Crypthecodinium cohnii*	?	Allen (1975)
Cyanochloronta	*Agmenellum quadruplicatum*	?	Roberts et al. (1975)

MV: variation in chromosome numbers

(Table 3.3). Because in comparison to land plants, most of the numbers have been assumed on the basis of morphology (Wyatt and Anderson 1984; Wyatt and Stoneburner 1984). Hybridization has been reported only in some cases (Smith 1978; Anderson 1980; Wyatt and Anderson 1984; Wyatt and Stoneburner 1984).

Table 3.3 Prevalence of polyploidy in bryophytes

Taxon	Chromosome numbers (*n*)
Amblystegiaceae	
Amblystegium serpens	7,9,10,11,12,14,19,20,21,22,24,40,48
Amblystegium varium	10,11,13,20,21, 40
Cratoneuron filicinum	9,10,11,20,30
Leptodictyum riparium	10,12,20,21,24,36,40,72
Bryaceae	
Bryum angustirete	10, 20, 27, 30
Bryum creberrinum	10, 22, 30
Bryum inclinatum	10, 20, 30
Bryum pseudotriquetrum	10, 11, 12, 20, 22, 33
Leptobryum pyriforme	10, 20, 21, 22, 24, 33
Pohlia cruda	10, 11, 14, 22, 40
Pohlia nutans	11, 20, 21, 22, ~27, 33
Funariaceae	
Funaria hygrometrica	14, 21, 28, 56
Physcomitrium pyriforme	9, 18, 26, 27, 36, 45, 52, 54
Polytrichaceae	
Atrichum undulatum	7, 14, 21
Polytrichum piliferum	7, 14, 21
Pottiaceae	
Phascum cuspidatum	21, 26, 27, 28, 42, 52
Tortula mucronifolia	12, 13, 24, 26, 30, 52,
Tortula muralis	13, 14, ~23, 24, 26, 27, ~28, 30, ~39, ~40, 48, 50, 52, 55, 60, 66
Tortula princeps	12, 24, 25, 26, 28, 38
Tortula subulata	12, 13, 14, 24, 26, 48, 49, ~60

Adapted from Elzbieta Kuta and Leslaw Przywara Acta Biologica Cracoviensia Series Botanica 39: 17–26, 1997. Polyploidy in Mosses

3.4 Polyploidy in Pteridophytes

Among vascular plants, polyploidy occurs in all the major divisions, viz. Psilopsida, Lycopsida, Sphenopsida, Filicopsida, gymnosperms and angiosperms (Stebbins 1950). Very high chromosome numbers, suggestive of polyploid series, have been found in both the living genera of Psilopsida (*Psilotum* and *Tmesipteris*) and in nearly all the genera of Filicopsida (ferns), especially Polypodiaceae. Levels of polyploidy in pteridophytes are said to reach the highest known in the plant kingdom, with some estimates as high as 95% (Grant 1981). *Ophioglossum reticulatum* with 1440 chromosomes is thought to be 96-ploid (Khandelwal 1990). On the basis of distribution of gametic chromosome numbers, Otto and Whitton (2000) estimated that approximately 7% of all speciation events in ferns involve polyploids.

3.5 Polyploidy in Gymnosperms

In living gymnosperms, the occurrence of polyploidy is particularly uncommon (Khoshoo 1959; Murray 1998). It is unknown in cycads and *Ginkgo*; among conifers known only are some species of *Pseudolarix* (Sax and Sax 1933), *Sequoia* (Dark 1932; Buchholz 1939a, b), *Juniperus* (Sax and Sax, 1933) and *Podocarpus* (Flory 1936). Tetraploidy has been reported in *Gnetum*, *Welwitschia* and *Ephedra* (Florin 1932), but these reports have been considered doubtful in the first two genera (Delevoryas 1980). Indeed, there are very few reports of polyploidy in gymnosperms (less than 5% of species). Khoshoo (1959) listed only 11 real polyploids in gymnosperms out of 240 (4.6%) taxa investigated (Table 3.4).

3.6 Polyploidy in Angiosperms

Among the angiosperms, polyploidy is very common. Many estimates of the frequency of polyploidy in flowering plants have been made by using chromosome numbers as dividing line between the diploids and polyploids. These estimates have varied depending on the base chromosome numbers used, as well as on the sample of taxa considered (Soltis and Soltis 2009). Various authors have estimated the frequency of polyploidy ranging from 30–70%. Müntzing (1936) and Darlington (1937) reported that about one half of all angiosperm species were polyploids, while Stebbins (1950) estimated the frequency of polyploidy in angiosperms to be 30–35%. Grant (1963, 1981), based on the assumption that the ancestral number of chromosomes in angiosperms was 7–9, estimated that any flowering plant with $n = 14$ or higher had undergone polyploidization at some point during the angiosperm evolution. Using this cut-off point, Grant inferred that 47% of all flowering plants were of polyploid origin, the monocots comprising 58% and the dicots 43% polyploids. Goldblatt (1980) and Masterson (1994) argued that Grant's (1963) assumption of ancestral chromosome number in angiosperms was too conservative, suggesting instead that $n = 9$ or 10 was more appropriate; based on this postulation, they concluded that 70% of angiosperms are polyploids. Comparison of leaf guard cell size in fossil and extant taxa of a few angiosperm families (Platanaceae,

Table 3.4 Prevalence of polyploidy in gymnosperms

Species	Polyploid chromosome number	References
Pinus densiflora ($2n = 24$)	48	Zinnai and Chiba (1951)
P. radiata ($2n = 24$)	48	Rodger (1953–1954)
Picea abies ($2n = 24$)	±28,30, 36,48,	Kiellander (1950), Illies (1953)
Cryptomeria japonica ($2n = 22$)	33, 44	Chiba (1950) and Zinnai and Chiba (1951)
Welwitschia mirabilis ($2n = 42$)	84	Fernandes (1936)

Adapted from Khoshoo TN (1959)

Lauraceae, Magnoliaceae) has been used for estimating the occurrence of polyploidy through time by Masterson (1994). Based on these comparisons of guard cell size, which is often much larger in polyploids than in diploids, the author could guess whether the fossil taxa were diploid or polyploid and also that 70% of all angiosperms had experienced one or more episodes of polyploidy in their ancestry. On the basis of distribution of gametic chromosome numbers, Otto and Whitton (2000) estimated that roughly 2–4% of all speciation events in angiosperms involve polyploids. According to Wood et al. (2009), 35% of angiosperms are polyploids, while 15% speciation events in them are associated with increase in ploidy level.

The systematic distribution of the polyploid species, however, is not uniform throughout the group of angiosperms. Stebbins 1950) did arrive at the following conclusions: Polygonaceae, Crassulaceae, Rosaceae, Malvaceae, Araliaceae, Poaceae and Iridaceae have a high frequency of polyploidy, whereas Fagaceae, Moraceae, Berberidaceae, Polemoniaceae and Cucurbitaceae are nearly or entirely devoid of polyploidy; most genera of angiosperms contain some polyploidy; on an average the highest percentages of polyploid species occur in perennial herbs, followed by annuals, and the woody plants. According to Grant (1981), polyploidy is common in the Rosaceae, Rubiaceae, Asteraceae, Iridaceae, Poaceae and many other families, relatively less common in families such as Caesalpiniaceae, Passifloraceae and Polemoniaceae, while in families such as Fagaceae and Berberidaceae, it is rare. Leitch and Bennett (1997) also maintain that polyploidy is highly variable between different angiosperm families. Stace (2000) holds that both polyploidy and interspecific hybridization are prevalent in Poaceae, while opposite is true for Fabaceae and Apiaceae. Raven et al. (2005) report an estimated 47% to more than 70% of flowering plants as being polyploid, as are also about 80% of the species in Poaceae.

Polyploids often occupy habitats different from those of their diploid parents and have been considered to be superior colonizers than their diploid competitors. Due to the presence of extra genomes, polyploids exhibit increased demand for nutrients, particularly nitrates and phosphates, that are needed to make nucleic acids and proteins (Leitch and Leitch 2008). This implies that occurrence of polyploidy will be less in nutrient-deficient areas. Some traits, such as increased levels of drought tolerance; effective means of vegetative reproduction; apomixis; pest resistance; variation in organ size, flowering time and biomass; and cross-fertilization (allogamy) may allow polyploids to enter new ecological niches or improve their fitness, thereby favouring establishment of polyploidy. Polyploids may survive better than their diploid progenitors in harsh environmental conditions, such as high altitudes and latitudes, and cold climates (Stebbins 1950, 1971; Grant 1981).

3.7 Polyploidy in Fungi

In fungi several species were identified as natural polyploids as shown in Table 3.5. *Rhizopus oryzae* was the first fungus to be fully sequenced. Subsequently it was analysed and was found to be polyploid followed by diploidization (Ma et al. 2009).

Table 3.5 Prevalence of polyploidy in fungi

Fungal polyploid species	Fungal group
Allomyces macrogynus	Blastocladiomycota
Allomyces arbuscula	
Allomyces javanicus	
Glomus etunicatum	Glomeromycota
Phyllactinia guttata	Ascomycota
Botrytis allii	
Stephensia shanorii	
Xylaria curta	
Saccharomyces cerevisiae	
Saccharomyces cariocanus	
Saccharomyces paradoxus	
Saccharomyces mikatae	
Saccharomyces kudriavzevii	
Saccharomyces bayanus	
Saccharomyces uvarum	
Saccharomyces pastorianus	
Zygosaccharomyces rouxii	
Dekkera bruxellensis	
Agaricus campestris	Basidiomycota
Agaricus bisporus	
Flammulina sp.	
Cyathus stercoreus	
Coprinellus micaceus	
Coprinopsis atramentaria	
Coniophora puteana	
Microbotryum violaceum	
Melampsora medusae-populina	
Phycomyces blakesleeanus	Mucoromycotina
Rhizopus oryzae	

Whole genome duplication has been recently identified within the *Blastocladiomycota* phylum particularly in aquatic fungi *Allomyces* sp. It consists of autotriploids, autotetraploids and allotetraploid series (Emerson and Wilson 1954). In *Glomus etunicatum* it has been reported that polyploidization might help for successful establishment in the absence of sexual reproduction (Pawlowska and Taylor 2004). Polyploidy has also been reported in *Basidiomycota* phylum. In *Ascomycota*, the largest phylum of fungi, polyploidy is also present. Most of the evidences regarding polyploidy in fungi have been reported in *Saccharomyces* genus. After the divergence of *Saccharomyces* from *Kluyveromyces*, the WGD take part around 100 Ma. Several studies revealed that a large number of yeasts are recent polyploids, in addition to ancient polyploids (Naumov et al. 2000). A large number of food-processing yeasts have been reported to be autotetraploids (Albertin et al. 2009). Many yeast strains used for wine processing have been found to be polyploid.

3.8 Occurrence of Autopolyploidy

Allopolyploids are considered to be more common than autopolyploids, but importance is given to autopolyploids (Soltis and Soltis 1993; Soltis et al. 2004; Tate et al. 2005; Wendel and Doyle 2005; Soltis et al. 2007). In natural autopolyploids, irregular chromosome pairing at meiosis results in sterility because multivalent formation takes place, which can lead to reduced fertility. However, stability of chromosome pairing in some even numbered polyploids does not stop polyploid to reproduce. In newly formed wheat polyploids, rapid sequence elimination revealed that differential loss genome-specific sequences may facilitate pairing of homologous chromosomes (Shaked et al. 2001). Also, it has been suggested that autopolyploids may shift from multivalent formation and polysomic inheritance to disomic inheritance (Le Comber et al. 2010). It is because polysomic inheritance has been recognized as a major criterion for detection of autopolyploid (Muller 1914; Haldane 1930).

Molecular markers such as allozymes and microsatellites have revolutionized the study of autopolyploidy as disomic/polysomic inheritance is concerned (Tate et al. 2005; Stift et al. 2008). Based on allozyme data, several autopolyploid species have been recognized, which earlier were unrecognized, including *Turnera ulmifolia*, *Allium nevii*, *Tolmiea menziesii*, *Heuchera micrantha* and *H. grossulariifolia* (Soltis and Soltis 1993, 1999; Tate et al. 2005). The updated list of possible autopolyploids is given in Table 3.6 (adapted from Soltis et al. 2014). There are many studies which suggested that autopolyploids are frequently generated, but many either remains undetected or quickly lost (Ramsey and Schemske 1998). Ramsey and Schemske (1998) give an estimation that the rate of polyploid formation is high as compared to mutation rate. The cytotypes recognized as intraspecific chromosomal process are mostly autopolyploids. Autopolyploidy may be more prevalent in some plant groups (Saxifragaceae and Cactaceae) than others (Soltis and Soltis 1993; Hamrick et al. 2002; Arakaki et al. 2007; Majure et al. 2012). In mosses most known polyploids appear to be autopolyploids (Husband et al. 2013). As reviewed by Soltis et al. (2014), among 2647 species from 346 genera belonging to 62 angiosperm families, around 334 species (13%) have multiple cytotypes. Majority of these cytotypes are thought to be autopolyploids but will require thorough and careful study.

There are several reasons for success of autopolyploids in nature including polysomic inheritance (Soltis and Soltis 1989). As a result heterozygosity increases in autopolyploids as compared to diploids, and four different alleles are found per locus in an autopolyploid plant (Moody et al. 1993; Soltis and Soltis 1993). During the first few generations of genome doubling, autopolyploids do not undergo major genomic changes. However, eventually after a long period of time, these changes took place (Parisod et al. 2010).

Biogeographic and niche modelling analysis revealed that many autopolyploids have established due to early divergence in habitat requirements. Some studies suggest that after formation of autopolyploids, they occupy novel niche space (Ramsey 2011; Levin 2011). Autotetraploid and diploid cytotypes of *Heuchera cylindrical*

Table 3.6 Recognized autopolyploids and their mode of origin

S No.	Taxon	Ploidy	Type of evidence
1	*Acacia nilotica*	4 *x*	Segregation pattern
2	*Actinidia chinensis* var. *deliciosa*	4 *x*	Segregation pattern and meiotic associations
3	*Adansonia digitata*	4 *x*	Segregation pattern
4	*Allium nevii*	4 *x*	Segregation pattern
5	*Arabidopsis lyrata*	4 *x*	Segregation pattern
6	*Aster amellus*	6 *x*	Morphology and lack of fixed heterozygosity
7	*Biscutella laevigata*	4 *x*	Segregation pattern
8	*Centaurea jacea*	4 *x*	Segregation pattern
9	*Chamerion angustifolium*	4 *x*	Segregation pattern
10	*Chrysanthemum morifolium*	6 *x*	Morphological segregation
11	*Dactylis glomerata*	4 *x*	Segregation pattern
12	*Dahlia variabilis*	4 *x*	Morphology
13	*Dioscorea alata*	4 *x*	Segregation pattern
14	*Dioscorea trifida*	4 *x*	Segregation patterns
15	*Galax urceolata*	4 *x*	Chemistry and lack of fixed heterozygosity
16	*Gallium anisophyllus*	4 *x*	Morphology
17	*Haplopappus spinulosus*	4 *x*	Segregation pattern
18	*Heuchera cylindrica*	4 *x*	Segregation pattern
19	*Heuchera grossulariifolia*	4 *x*	Segregation pattern
20	*Heuchera micrantha*	4 *x*	Segregation pattern
21	*Lotus corniculatus*	4 *x*	Segregation pattern
22	*Lythrum salicaria*	4 *x*	Morphological segregation
23	*Maclura pomifera*	4 *x*	Segregation pattern
24	*Medicago falcata*	4 *x*	Morphology and segregation pattern
25	*Medicago sativa*	4 *x*	Segregation pattern
26	*Pachycereus pringlei*	4 *x*	Segregation pattern
27	*Panicum virgatum*	8 *x*	Segregation pattern
28	*Paspalum notatum*	4 *x*	Segregation pattern
29	*Phleum pratense*	6 *x*	Morphology
30	*Prunus spinosa*	4 *x*	Segregation pattern
31	*Rorippa amphibia*	4 *x*	Segregation pattern
32	*Rorippa sylvestris*	4 *x*	Segregation pattern
33	*Rutidosis leptorrhynchoides*	4 *x*	Segregation pattern
34	*Solanum tuberosum*	4 *x*	Morphology and segregation pattern
35	*Thymus praecox*	4 *x*	Segregation pattern
36	*Tolmiea menziesii*	4 *x*	Segregation pattern
37	*Turnera sidoides*	Complex 2 *x*–8 *x*	Meiotic associations
38	*Turnera ulmifolia* var. *elegans*	4 *x*	Segregation pattern

(continued)

Table 3.6 (continued)

S No.	Taxon	Ploidy	Type of evidence
39	*Turnera ulmifolia* var. *intermedia*	4 *x*	Morphology and segregation pattern
40	*Vaccinium corymbosum*	4 *x*	Segregation pattern

Adapted from: Soltis et al. (2014). The polyploidy revolution then and now: Stebbins revisited. Am J Bot 101: 1057–1078)

both grow in different ranges, but their niche requirements have not being diverged (Godsoe et al. 2013). Autotetraploid *Tolmiea menziesii* appears to be distinct from its progenitor, *T. diplomenziesii*, both geographically and ecologically (Soltis 1984). In *Houstonia*, divergence between diploid and tetraploid cytotypes was found in *H. longifolia*, but divergence was not observed in *H. purpurea* (Glennon et al. 2012).

References

Albertin W, Marullo P, Aigle M, Bourgais A, Bely M, Dillmann C, De Vienne D, Sicard D (2009) Evidence for autotetraploidy associated with reproductive isolation in *Saccharomyces cerevisiae*: towards a new domesticated species. J Evol Biol 22:2157–2170

Allen JR (1975) Characterization of the DNA from dinoflagellate *Crypthecodinium cohnii* and implications for nuclear organization. Cell 6:161–169

Anderson LE (1980) Cytology and reproductive biology of mosses. In: Taylor RJ, Leviton AE (eds) The Mosses of North America-based on symposium on the biology of mosses. AAAS Pacific Division, San Francisco, pp 37–76

Arakaki M, Soltis DE, Speranza P (2007) New chromosome counts and evidence of polyploidy in *Haageocereus* and related genera in tribe Trichocereeae and other tribes of Cactaceae. Brittonia 59:290–297

Brandham PE (1965) Polyploidy in desmids. Can J Bot 43:405–417

Brandham PE, Godward MBE (1964) The production and inheritance of haploid triradiate form in *Cosmarium botrytis*. Phycologia 4:75–83

Buchholz JT (1939a) The embryogeny of *Sequoia sempervirens* with a comparison of sequoias. Am J Bot 26:248–257

Buchholz JT (1939b) Generic segregation of the sequoias. Am J Bot 26:535–538

Buffaloe NP (1958) A comparative cytological study of four species of *Chlamydomonas*. Bull Torrey Bot Club 85:157–178

Cave MS, Pocock MA (1956) Variable chromosome number in *Astrephomene guberniculifera*. Am J Bot 43:122–134

Chiba S (1950) Triploids and tetraploids of Sugi (*Cryptomeria japonica* D. Don.) selected in the forest nursery. Bull Govt For Expt Sta 49:99–108

Dark SOS (1932) Chromosomes of Taxus, Sequoia, Cryptomeria and Thuya. Ann Bot 46:965–977

Darlington CD (1937) Recent advances in cytology, 2nd edn. Churchill, London

Delevoryas T (1980) Polyploidy in gymnosperms. In: Polyploidy. Springer, Berlin, pp 215–218

Emerson R, Wilson CM (1954) Interspecific hybrids and the cytogenetics and cytotaxonomy of *Euallomyces*. Mycologia 46:393–434

Fankhauser G (1945) Maintenance of normal structure in heteroploid salamander larvae through compensation of changes in cell size by adjustment of cell number and cell shape. J Exp Zool 100:445–455

Fernandes A (1936) Sur la caryologie de *Welwitschia mirabilis* Hook. f. Bol Soc Brot 11:267–282

Florin R (1932) Die Chromosomenzahlen bei Welwitschia und einigen Ephedra Arten. Svensk Bot Tidsk 26:205–214
Flory WS (1936) Chromosome numbers and phylogeny in gymnosperms. J Arnold Arbor 17:83–89
Godsoe W, Larson MA, Glennon KL, Segraves KA (2013) Polyploidization in *Heuchera cylindrica* (Saxifragaceae) did not result in a shift in climatic requirements. Am J Bot 100:496–508
Godward MBE (1950) On the nucleolus and nucleolar organizing chromosomes of Spirogyra. Ann Bot (Lond) NS 14:39–53
Godward MBE (1966) The chromosomes of the algae. Edward Arnold, London
Goldblatt P (1980) Polyploidy in angiosperms: monocotyledons. In: Lewis WH (ed) *Polyploidy: biological relevance*. Plenum Press, New York, pp 219–239
Goldstein M (1964) Speciation and mating behaviour in Eudorina. J Protozool 11:317–344
Grant V (1963) *The origins of adaptations*. Columbia University Press, New York
Grant V (1981) Plant speciation, 2nd edn. Columbia University Press, New York
Haldane JBS (1930) Theoretical genetics of autopolyploids. J Genet 22:359–372
Hamrick JL, Nason JD, Fleming TH, Nassar JM (2002) Genetic diversity in columnar cacti. In: Fleming TH, Valiente-Banet A (eds) Columnar cacti and their mutualists: evolution, ecology, and conservation. University of Arizona Press, Tuscon, pp 122–133
Husband BC, Baldwin SJ, Suda J (2013) The incidence of polyploidy in natural plant populations: major patterns and evolutionary processes. In: Greilhuber J, Dolezel J, Wendel JF (eds) Plant genome diversity. Springer, Vienna, pp 255–276
Illies ZM (1953) (selection of multiple seedlings in single plant progenies of *Picea abies*.) Zeit Forstgen. Forstpflanz 2: 21. Vide P1. Breed Abstr 1953, 23: No. 1486
Khandelwal S (1990) Chromosome evolution in the genus Ophioglossum L. Bot J Linn Soc 102(3):205–217
Khoshoo TN (1959) Polyploidy in gymnosperms. Evolution 13:24–39
Kiellander CL (1950) Polyploidy in Picea abies. Hereditas 36:513–516
Glennon K, Rissler LJ, Church SA (2012) Ecogeographic isolation: a reproductive barrier between species and between cytotypes in *Houstonia* (Rubiaceae). Evol Ecol 26:909–926
Kuta E, Przywara L (1997) Polyploidy in mosses. Acta Biol Cracov Bot 39:17–26
Le comber SC, Ainouche ML, Kovarik A, Leitch AR (2010) Making a functional diploid: from polysomic to disomic inheritance. New Phytol 186:113–122
Leitch IL, Bennett MD (1997) Polyploidy in angiosperms. Trends Plant Sci 2:470–476
Leitch AR, Leitch IJ (2008) Genomic plasticity and the diversity of polyploid plants. Science 320:481–483
Levin D (2011) Polyploidy and ecological transfiguration in *Achillea*. Proc Nat Acad Sci U S A 108:6697–6698
Lewis WH (1980) Polyploidy in species populations. In: Lewis WH (ed) *Polyploidy: biological relevance*. Plenum, New York, pp 103–144
Ling HV, Tyler PA (1976) Meiosis, polyploidy and taxonomy of the *Pleurotaenium mamillatum* complex (desmidiaceae). Br Phycol J 11:315–330
Ma L-J et al (2009) Genomic analysis of the basal lineage fungus *Rhizopus oryzae* reveals a whole-genome duplication. PLoS Genet 5:e1000549
Majure LC, Judd WS, Soltis PS, Soltis DE (2012) Cytogeography of the *Humifusa* clade of *Opuntia* s.s. Mill. 1754 (Cactaceae, Opuntioideae, Opuntieae): correlations with Pleistocene refugia and morphological traits in a polyploid complex. Comp Cytogenet 6:53–77
Masterson J (1994) Stomatal size in fossil plants: evidence for polyploidy in majority of angiosperms. Science 264:421–424
Moody ME, Mueller LD, Soltis DE (1993) Genetic variation and random drift in autotetraploid populations. Genetics 154:649–657
Muller HJ (1914) A new mode of segregation in Gregory's tetraploid primulas. Am Nat 48:508–512
Muller DG (1976) Generation suechsel, Kernphasenwechsel und Sexualitat der *Ectocarpus siliculosus* im kolturversuch. Planta 75:39–54
Müntzing A (1936) The evolutionary significance of autopolyploidy. Hereditas 21:263–378
Murray BG (1998) Nuclear DNA amounts in gymnosperms. Ann Bot 82(Suppl 1):3–15

Naumov GI, Naumova ES, Masneuf I, Aigle M, Kondratieva VI, Dubourdieu D (2000) Natural polyploidization of some cultured yeast Saccharomyces sensu stricto: auto- and allotetraploidy. Syst Appl Microbiol 23:442–449

Newton ME (1984) The cytogenetics of bryophytes. In: Dyer AF, Duckett JG (eds) The experimental biology of bryophytes. Academic, London, pp 65–96

Nichols HW (1980) Polyploidy in algae. In: Lewis WH (ed) Polyploidy: biological relevance. Plenum Press, New York, pp 151–161

Otto SP, Whitton J (2000) Polyploid incidence and evolution. Annu Rev Genet 34:401–437

Parisod C, Holderegger R, Brochmann C (2010) Evolutionary consequences of autopolyploidy. New Phytol 186:5–17

Pawlowska TE, Taylor JW (2004) Organization of genetic variation in individuals of arbuscular mycorrhizal fungi. Nature 427:733–737

Ramsey J (2011) Polyploidy and ecological adaptation in wild yarrow. Proc Natl Acad Sci USA 108:7096–7101

Ramsey J, Schemske DW (1998) Pathways, mechanisms, and rates of polyploid formation in flowering plants. Annu Rev Ecol Syst 29:467–501

Rao BGS, Mantha S, Rao MU (1977) Chromosome behaviour at meiosis and its bearing on the cytotaxonomy of *Ceramium* species. Bot Marina 21:123–129

Raven PH, Evert RF, Eichhorn SE (2005) Biology of plants, 7th edn. W. H Freeman, New York

Roberts TM, Loeblich R, Klotz C (1975) Studies on the DNA of the blue green alga *Agmenellum quadruplicatum*. J Phycol 11 (suppl.):16

Rodger GJ (1953–1954) Annual report of the forestry and timber bureau, commonwealth of Australia for 1953–1954. Vide Pl breed Abstr 1956, 26:2861

Sax K, Sax HJ (1933) Chromosome number and morphology in conifers. J Arnold Arbor 14:356–375

Shaked H, Kashkush K, Ozkan H, Feldman M, Levy AA (2001) Sequence elimination and cytosine methylation are rapid and reproducible responses of the genome to wide hybridization and allopolyploidy in Wheat. Plant Cell 13:1749–1759

Sinha JP (1958) Cytological and cultural study of some members of Cladophorales and Oedogoniales, Ph.D. thesis, London University

Smith AJE (1978) Cytogenetics, biosystematics and evolution in Bryophyta. Adv Bot Res 6:195–276

Soltis DE (1984) Autopolyploidy in *Tolmiea menziesii* (Saxifragaceae). Am J Bot 71:1171–1174

Soltis DE, Soltis PS (1989) Genetic consequences of autopolyploidy in *Tolmiea* (Saxifragaceae). Evolution 43:586–594

Soltis DE, Soltis PS (1993) Molecular data facilitate a reevaluation of traditional tenets of polyploidy evolution. Crit Rev Plant Sci 12:243–273

Soltis DE, Soltis PS (1999) Polyploidy: recurrent formation and genome evolution. Trends Ecol Evol 14:348–352

Soltis PS, Soltis DE (2009) The role of hybridization in plant speciation. Annu Rev Plant Biol 60:561–588

Soltis DE, Soltis PS, Pires JC, Kovarik A, Tate J, Mavrodiev E (2004) Recent and recurrent polyploidy in *Tragopogon* (Asteraceae): cytogenetic, genomic and genetic comparisons. Biol J Linn Soc 82:485–501

Soltis DE, Soltis PS, Schemske DW, Hancock J, Thompson J et al (2007) Autopolyploidy in angiosperms: have we grossly underestimated the number of species? Taxon 56:13–30

Soltis DE, Visger CJ, Soltis PS (2014) The polyploidy revolution then…and now: stebbins revisited. Am J Bot 101:1057–1078

Stace CA (2000) Cytology and cytogenetics as a fundamental taxonomic resource for the 20th and 21st centuries. Taxon 49:451–477

Starr RC (1958) The production and inheritance of the triradiate form in *Cosmarium turpinii*. Am J Bot 45:243–248

Stebbins GL (1938) Cytological characteristics associated with the different growth habits in the dicotyledons. Am J Bot 25:189–198

Stebbins GL (1950) Variation and evolution in plants. Columbia University Press, New York
Stebbins GL (1971) Chromosomal evolution in higher plants. Addison-Wesley, London
Stift M, Berenos C, Kuperus P, Van Tienderen PH (2008) Segregation models for disomic, tetrasomic and intermediate inheritance in tetraploids: a general procedure applied to *Rorippa* (yellow cress) microsatellite data. Genetics 179:2113–2123
Tate JA, Soltis DE, Soltis PS (2005) Polyploidy in plants. In: Gregory TR (ed) The evolution of the genome. Elsevier, San Diego, pp 371–426
Tischler G (1931) Pflanzliche Chromosomen-Zahlen. - Tabul. Biol 7:109–226
Tischler G (1938) Pflanzliche Chromosomen-Zahlen. Tabulae Biol 16:162–218
Voglmayr H (2000) Nuclear DNA amounts in mosses (Musci). Ann Bot 85(4):531–546
Wendel JF, Doyle JJ (2005) Polyploidy and evolution in plants. In: Henry RJ (ed) Plant diversity and evolution: genotypic and phenotypic variation in higher plants. CABI, Wallingford, pp 97–117
Wettstein FV (1927) Die Erscheinung der Heteroploidie, hesonders imPflanzenreich. – Ergebri. der Biol., Bd. 11. Springer, Berlin
Wood TE, Takebayashi N, Barker MS, Mayrose I, Greenspoon PB, Rieseberg LH (2009) The frequency of polyploid speciation in vascular plants. Proc Natl Acad Sci U S A 106:13875–13879
Wyatt R, Anderson LE (1984) Breeding systems of bryophytes. In: Dyer AF, Duckett JG (eds) The experimental biology of bryophytes. Academic, London, pp 39–64
Wyatt R, Stoneburner A (1984) Biosystematics of bryophytes: an overview. In: Grant WF (ed) Plant biosystematics. Academic, Toronto/Orlando/San Diego/New York/London/Montreal/San Francisco/Sydney/Tokyo/Sao Paulo, pp 519–542
Zinnai I, Chiba S (1951) Naturally occurring tetraploids of *Cryptomeria japonica* Ikushugaku Zasshi. Jap J Breed 1:43–46. Vide P1. Breed Abstr 1952, 22: No 1473

Detection of Polyploidy 4

4.1 Criteria for Detection

Earlier, polyploidy was detected on the basis of only a few morphological, cytological and genetic criteria. However, it may not be correct to characterize polyploidy on the above structural basis only because the chromosome pairing behaviour is also affected by other factors (Otto 2007). Polyploids have distinct morphology as compared to their diploid counterparts. Increased cell size that results in thicker and broader leaves and larger flowers and fruits is often associated with polyploidy. These characters are often used to identify putative polyploids, when screening of large number of plants has to be done. Larger pollen size and stomata are also often used to detect polyploidy. Although Stebbins (1950) had successfully used morphological criteria as evidence for finding the progenitor of the B-genome, yet some controversy still exists. Besides morphology, biochemical compounds, such as secondary metabolites, seed proteins and isozymes, have been also used to investigate progenitor of allopolyploids.

4.2 Cytological Estimation

Cytological studies of mitosis and meiosis can be used to detect polyploidy. For instance, autopolyploid nature of a polyploid can be confirmed if more than two sets (three in triploids and four in tetraploids) of morphologically similar chromosomes are seen in karyotypes. If there are only two sets of similar chromosomes, then a comparison can be made between the karyotype of polyploid and those of the possible diploid progenitors to find out the origin of such a polyploid. Meiotic studies are also used for establishing the nature of polyploids and for conducting genome analysis. For example, in an autopolyploid, one expects the presence of mainly multivalents, and in an allopolyploid only bivalents are present, whereas in a segmental allopolyploid, both bivalents and multivalents are formed in variable numbers (Gupta 2007). Polysomic segregation ratios have also been used to distinguish

T.-Ul.-H. Dar, R.-Ul. Rehman, *Polyploidy: Recent Trends and Future Perspectives*, https://doi.org/10.1007/978-81-322-3772-3_4

autopolyploids from allopolyploids, the former exhibiting polysomic inheritance and the latter only disomic inheritance.

4.3 Nuclear DNA Estimation

Estimation of nuclear DNA content is another approach to detect polyploidy. Of the various methods used to estimate genome size such as Feulgen microdensitometer, fluorometry, reassociation kinetics and biochemical analysis, flow cytometry is the most widely used in plants (Bennett and Leitch 2005). Polyploid complexes have created problems in recognizing plant taxa; by using flow cytometry, this problem has been resolved in a number of vascular plants with ploidy variation (reviewed in Kron et al. 2007). Flow cytometry has been used to identify diploid progenitors by comparing genome size of polyploids to those of their putative diploid progenitors (Bennett and Leitch 2005). Besnard et al. (2008), based on flow cytometry, confirm polyploidy in the olive complex (*Olea europaea*).

4.3.1 Flow Cytometry

The different steps involved in flow cytometry include isolation of nuclei from leaf tissue by chopping and, then, staining of isolated nuclei with fluorescent dye that binds quantitatively to DNA and then passed through a flow cytometry. The nuclei then passed one at a time through a series of lights, lenses, mirrors and amplifiers that detect and convert the amount of fluorescent light being emitted by each nucleus into a digital signal. The absolute DNA content of the plant can be determined by comparing the intensity of fluorescence with that from a plant of a known DNA amount (Fig. 4.1).

4.3.2 Feulgen Microdensitometer

This method involves staining the nuclei and then measuring the amount of light observed by the stain. Feulgen staining is the most commonly used method of staining and was first developed by Feulgen and Rossenbeck in 1924. In this method, acid hydrolysis is performed which frees aldehyde groups of DNA. This is followed by staining with basic fuchsin (Schiff's reagent); as a result purple colour is formed when it complexes with the aldehyde groups. For Feulgen microdensitometer, plant material is prepared using various modifications (McLeish and Sunderland 1961; Leuchtenberger 1958). Feulgen staining is specific and stoichiometric for DNA, the accurate estimation of C-value is provided by measuring total nuclear stain concentration compared against a known DNA amount (standard) (Bennett and Leitch 2005).

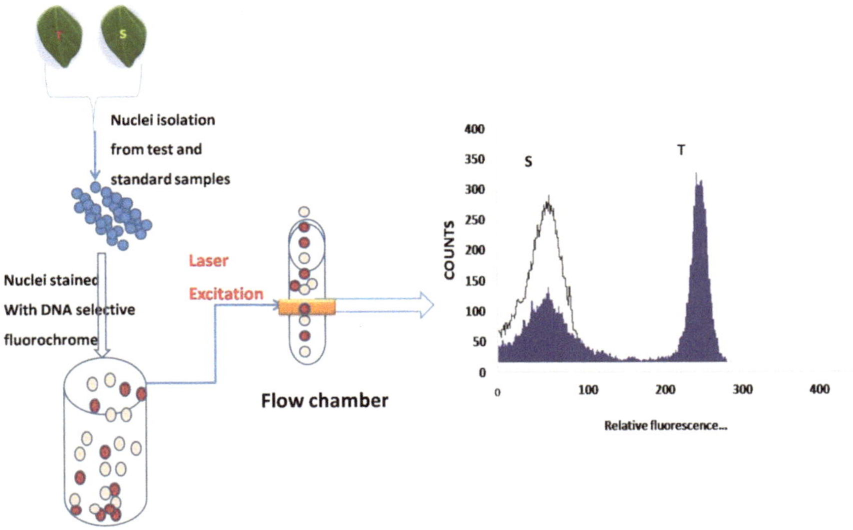

Fig. 4.1 Diagrammatic representation of flow cytometry. Cells or nuclei are extracted and stained with a DNA-specific dye. These are then passed through a laser beam. The fluorescence is measured and plotted in the form of histogram or scatterplot. The fluorescence of tested sample (T) is compared to that of standard (S) (Modified from Kron et al. 2007)

4.3.3 Reassociation Kinetics

In this method, DNA is extracted from cells and then denatured by heating. The rate of reassociation of the DNA strand indicates the relative copy of repeated DNA sequences. This can be followed by calibration against standard in order to give an estimate of absolute DNA content. This method of detecting polyploidy is very slow and not particularly accurate.

4.3.4 Biochemical Analysis

For the detection of polyploidy, different biochemical techniques such as differences in the distribution of secondary metabolites, some species-specific enzyme markers, seed proteins, DNA content, heterochromatin content, etc. can also be used. For example, in allopolyploids the content of phenolic compounds, etc. can be compared with diploid species.

4.4 Molecular Cytogenetics

Nowadays molecular cytogenetic techniques, such as fluorescent in situ hybridization (FISH) and genome in situ hybridization (GISH), are widely used to analyse genome organization of various polyploid taxa. GISH technique has the advantage of

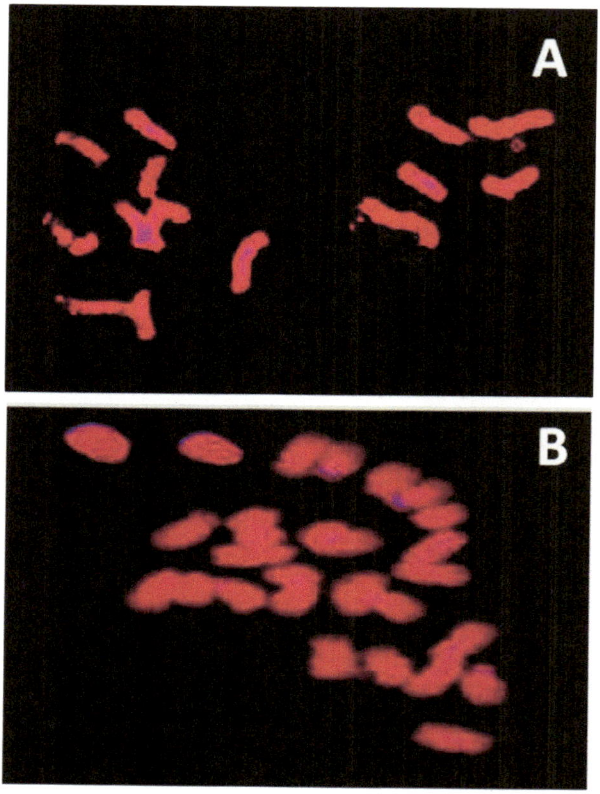

Fig. 4.2 (**a**) Metaphase chromosome preparation of diploid ($2n = 2x = 14$) Phlox drummondii probed with Dig-Rhodamine-labelled genomic DNA of colchitetraploid. (**b**) Metaphase chromosome preparation of colchitetraploid ($2n = 4x = 28$) Phlox drummondii probed with Dig-Rhodamine-labelled genomic DNA of diploid.

highlighting the whole chromosomes or genomes and is applicable to both dividing cells and those at interphase (Stace 2000). It has revolutionized the methods of identification and characterization of polyploid genomes. Bennett et al. (1992) for the first time used GISH for analysing a wild, naturally occurring polyploid, *Milliummontianum*. GISH was also used to observe species-specific inter-genomic translocations in allotetraploid tobacco which combines two ancestral genomes, known as the S and T genomes (Kenton et al. 1993). Multiple origins of the recently formed polyploid species of *Tragopogon* and *Spartina* have been confirmed by FISH and GISH (reviewed in Chester et al. 2010). On the bases of FISH and GISH analyses, Hasterok et al. (2004) confirmed that *Brachypodium distachyon* is an allopolyploid, not an autopolyploid as considered earlier. In autotetraploid *Phlox drummondii* by using GISH, no genomic changes were observed (Dar et al. unpublished) (Fig. 4.2).

4.4.1 FISH and GISH

Genomic in situ hybridization (GISH) is a modification of fluorescent in situ hybridization (FISH). This technique allows chromosomes from different parents/genomes in hybrid plants to be painted in different colours. Total genomic DNA from one

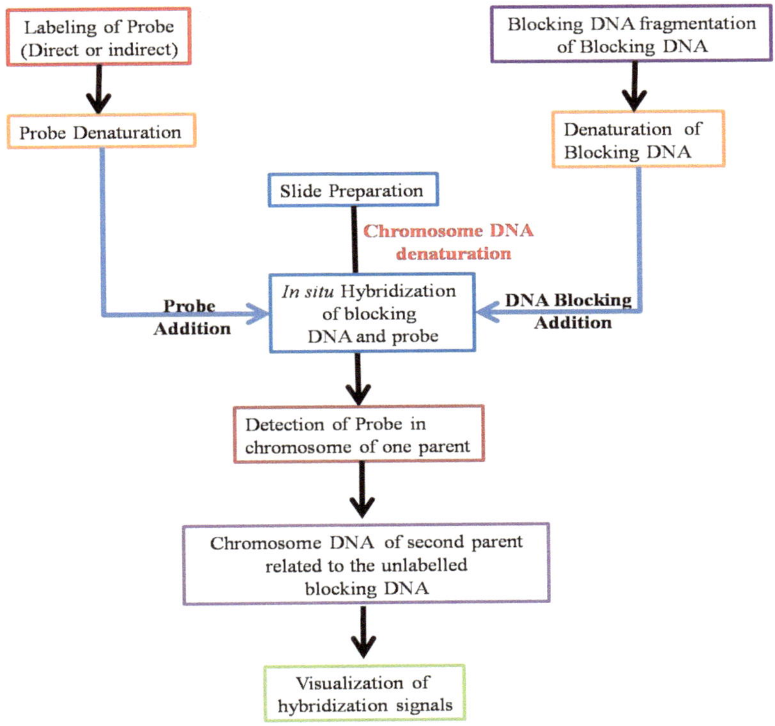

Fig. 4.3 Flow chart showing different steps involved in GISH

species is labelled as a probe and is hybridized to chromosomal DNA of other species in situ. The cells are then viewed under fluorescent microscope. Labelled probe that hybridizes to target DNA fluoresces differentially to the non-hybridized unrelated DNA sequences in the chromosomes (Fig. 4.3). GISH is a powerful tool for identification of component genome in allopolyploids (Raina and Raini 2001).

In *Tragopogon* species, fluorescent in situ hybridization (FISH) has been used to investigate possible chromosomal repatterning (Pires et al. 2004). Five different tandem repeats were used as probe on chromosomes of *Tragopogon*. Among the five, three (18S – 2 repeat 6S rDNA, 5S rDNA and telomeric sequence (TRS)) are commonly used in molecular cytogenetic studies. The other two (TPRMBO and TGP7) were isolated from *T. pratensis* and *T. porrifolius*, respectively. Three diploid progenitors (*T. dubius*, *T. pratensis* and *T. porrifolius*) of allotetraploid *Tragopogon* were analysed by FISH. These diploid species have six pairs of chromosomes. Allotetraploid species were also analysed by FISH using same probes in order to determine possible chromosome rearrangements after polyploidization. The distribution of tandem repetitive sequences was constructed in all the three diploid species of *Tragopogon* (Fig. 4.4). Thus, the location and intensity of FISH signals were ascertained for the identification of each diploid parental chromosome in polyploids. If after polyploidization, genomic rearrangements had taken place,

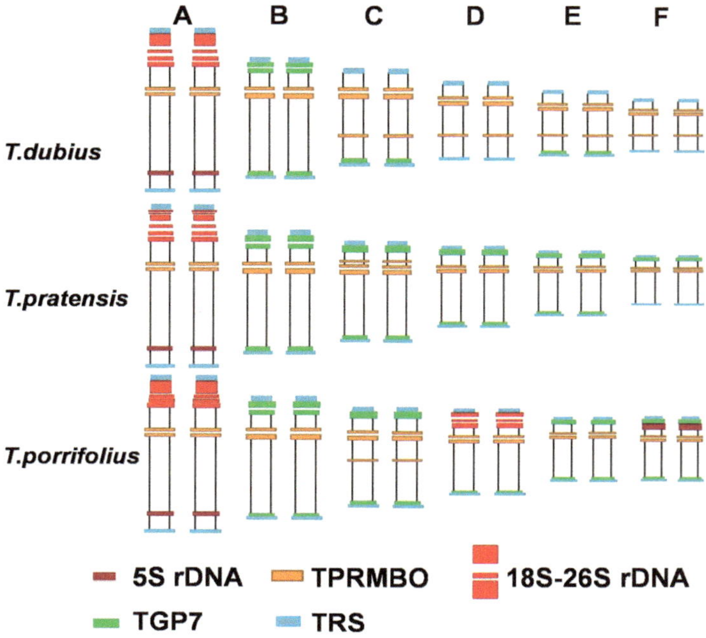

Fig. 4.4 Ideograms showing the distribution of five rDNA loci on chromosome of three diploid progenitors of allotetraploid Tragopogon by FISH. Six pairs of chromosomes are arranged from largest to smallest (A–F). In *T. porrifolius* as compared to other two diploid species, an extra 18S–26S rDNA locus on D and an extra 5S locus on F chromosome have been observed. Fewer TGP7 sub-telomeric repeats have been observed in *T. dubius* than other two diploid species. Secondary constructions are shown in gaps. They are found on short arms of A chromosome of all species and D chromosome of *T. porrifolius* (Modified from Pires et al. 2004)

then non-additive patterns would be expected. Higher or lower number of rDNA loci could be expected in polyploid than the sum of two diploid progenitors. No major genomic rearrangements have been observed in the allotetraploid species, *T. mirus* or *T. miscellus* (Pires et al. 2004).

4.5 Molecular Mapping

The molecular mapping of polyploids is problematic in comparison to diploids because the large populations are required for segregation analysis. Thus, the statistical analysis becomes complex for obtaining the reliable genetic distance (Porceddu et al. 2002). However for obtaining genetic distance, RFLP has been employed in *Triticum aestivum* (Liu et al. 1998), *Brassica napus* (Szadkowski et al. 2010; Gaeta et al. 2007; Nicolas et al. 2009; Cifuentes et al. 2010), *Draba norvegica* and *Tragopogon miscellus* (Leitch and Bennett 1997). Recently, AFLP has been employed in *Arabidopsis* (Madlung et al. 2005), *Brassica* (Song et al. 1995), *Spartina* (Salmon et al. 2005) and *Phlox* (Dar et al. 2013). Further, the autopolyploids are difficult to resolve, because of the fact that segregation in these depends on pairing behaviour of chromosomes (random vs preferential) and double reduction (Wu et al. 1992). A simple strategy adopted generally for mapping is to fuse single-dose markers from each parent (REF). Moreover, for polyploidy analysis, the statistical models have been developed for QTL mapping (Wu et al. 2004; Li et al. 2012). The chromosome pairing behaviour is estimated by molecular marker segregation. For instance, upon crossing autotetraploid genotypes of *Medicago sativa* subsp. *falcata* × *M. sativa* subsp. sativa, significant pairing was observed (Ma et al. 2002); however, in another study by Julier et al. (2003), it had no evidence of chromosome pairing in cultivated *M. sativa* genotypes. Furthermore, the study with Salix having allotetraploid origin supported the preferential chromosome pairing (Barcaccia et al. 2003). The study on the polysomic inheritance analysis proposed the crop improvements by determining number of allelic copies in a microsatellite locus (Esselink et al. 2004; Van Dijk et al. 2012). For instance, in allooctoploid strawberry (*Fragaria ananassa*), the microsatellite allele dose and configuration establishment (MADCE) was proposed for establishing the exact dose (Van Dijk et al. 2012). By using this technique, the applicability of SSR markers in polyploids gets enhanced as compared to the limitation of single-dose markers. Diversity Arrays Technology (DArT) has recently been used for mapping allotetraploid recombinant inbred lines obtained by crossing wild *Avena magna* ($2n = 4\times = 28$) with cultivated *A. sativa* ($2n = 6\times = 42$) (Oliver et al. 2011).

The whole genome duplication (WGD) results in high gene redundancy in the genomes of most of the higher eukaryotes. In *Arabidopsis*, 2–3 rounds of genome-wide duplication and numerous duplicate genes have been found by complete sequencing (Blanc et al. 2000; Vision et al. 2000; Bowers et al. 2003; Paterson et al. 2004). The nuclear genome sequencing in rice suggested ancient polyploidy (Paterson et al. 2004; Yu et al. 2005). Further on the basis of complete genome sequencing of *Vitis*, it was suggested that the common ancestor of *Vitis, Populus* and *Arabidopsis* was an ancient hexaploid that arose after the divergence of the monocot and eudicot lineages (Sampedro et al. 2005; Jaillon et al. 2007).

As revealed by genome mapping, the level of ploidy may have been underestimated for some species which were already recognized as polyploids (Leitch and Bennett 1997). For example, traditionally cotton has been considered to be allotetraploid ($2n = 4x = 52$). However, on the basis of RFLP map, Reinisch et al. (1994) confirmed it to be a paleo-octoploid, formed around 1.1–1.9 million years ago, following a polyploidization event that converted the basic chromosome number from 13 to 26. On the basis of RFLP analysis, Shoemaker et al. (1996) also reported chromosomal interchanges in soybean.

Ancient duplication events of plant genomes can be detected by EST sequences (Udall and Wendel 2006). The rapidly growing EST data sets of plants can be used to assess ancient polyploidy, using a genomics method to estimate gene duplication by various evolutionary processes, as well as the approximate age of these gene duplication events (Blanc and Wolfe 2004; Shi et al. 2010). Blanc and Wolfe (2004) analysed 14 model plant species for searching paralogs, or duplicated sequences, and found spikes in the distribution of older paralogous pairs. EST sequencing approach has been successfully used in a variety of plants, including *Arabidopsis* (Blanc and Wolfe 2004;Barker et al. 2009), wheat (Blanc and Wolfe 2004), maize (Blanc and Wolfe 2004; Schlueter et al. 2004), *Solanum* (Schlueter et al. 2004; Cui et al. 2006), *Populus* (Sterck et al. 2005), sunflower and lettuce (Barker et al. 2008).

References

Barcaccia G, Meneghetti S, Albertini E, Triest L, Lucchin M (2003) Linkage mapping intetraploid willows: segregation of molecular markers and estimation of linkage phases support an allotetraploid structure for *Salix alba* × *Salix fragilis* interspecific hybrids. Heredity 90:169–180

Barker MS, Kane NC, Matvienko M, Kozik A, Michelmore RW, Knapp SJ, Rieseberg LH (2008) Multiple paleopolyploidizations during the evolution of the Compositae reveal parallel patterns of duplicate gene retention after millions of years. Mol Biol Evol 25:2445–2455

Barker MS, Vogel H, Schranz ME (2009) Paleopolyploidy in the Brassicales: analyses of the Cleome transcriptome elucidate the history of genome duplications in *Arabidopsis* and other Brassicales. Genome Biol Evol 1:391–399

Bennett MD, Leitch IJ (2005) Nuclear DNA amounts in angiosperms: progress, problems and prospects. Ann Bot 95(1):45–90

Bennett ST, Kenton AY, Bennett MD (1992) Genomic *in situ* hybridization reveals the allopolyploid nature of *Milium montianum* (Gramineae). Chromosoma 101:420–424

Besnard G, Garcia-verdugo C, Rubiodecasas R, Treier UA, Galland N, Vargas P (2008) Polyploidy in the olive complex (Olea europaea): evidence from flow cytometry and nuclear microsatellite analyses. Annal Bot 101:25–30

Blanc G, Wolfe KH (2004) Functional divergence of duplicated genes formed by polyploidy during *Arabidopsis* evolution. Plant Cell 16:1679–1691

Blanc G, Barakat A, Guyot R, Cooke R, Delseny M (2000) Extensive duplication and reshuffling in the *Arabidopsis* genome. Plant Cell 12:1093–1101

Bowers JE, Chapman BA, Rong J, Paterson AH (2003) Unravelling angiosperm genome evolution by phylogenetic analysis of chromosomal duplication events. Nature 422:433–438

Chester M, Leitch AR, Soltis PS, Soltis DE (2010) Review of the application of modern cytogenetic methods (FISH/GISH) to the study of reticulation (polyploidy/hybridisation). Genes 1:166–192

Cifuentes M, Eber F, Lucas MO, Lode M, Chèvre AM, Jenczewski E (2010) Repeated polyploidy drove different levels of crossover suppression between homoeologous chromosomes in *Brassica napus* allohaploids. Plant Cell 22:2265–2276

Cui LY, Wall PK, Leebens-Mack JH, Lindsay BG, Soltis DE, Doyle JJ, Soltis PS, Carlson JE, Arumuganathan K, Barakat A, Albert VA, Ma H, DePamphills CW (2006) Widespread genome duplications throughout the history of flowering plants. Genome Res 16:738–749

Dar TH, Raina SN, Goel S (2013) Molecular analysis of genomic changes in synthetic autotetraploids *Phlox drummondii* Hook. Biol J Linn Soc 110:591–605

Esselink GD, Nybom H, Vosman B (2004) Assignment of allelic configuration in polyploids using the MAC-PR (microsatellite DNA allele counting—peak ratios) method. Theor Appl Genet 109:402–408

Gaeta RT, Pires JC, Iniguez-Luy F, Leon E, Osborn TC (2007) Genomic changes in resynthesized *Brassica napus* and their effect on gene expression and phenotype. Plant Cell 19:3403–3417

Gupta PK (2007) Cytogenetics. Rastogi Publications, Meerut

Hasterok R, Draper J, Jenkins G (2004) Laying the cytotaxonomic foundations of a new model grass, *Brachypodium distachyon* L. Chromosome Res 12:397–403

Jaillon O, Aury JM, Noel B, Policriti A, Clepet C et al (2007) French-Italian public consort. Grapevine genome charact. The grapevine genome sequence suggests ancestral hexaploidization in major angiosperm phyla. Nature 449:463–467

Julier B, Flajoulot S, Barre P, Cardinet G, Santoni S, Huguet T, Huyghe C (2003) Construction of two genetic linkage maps in cultivated tetraploid alfalfa (*Medicago sativa*) using microsatellite and AFLP markers. BMC Plant Biol. https://doi.org/10.1186/1471-2229-3-9

Kenton A, Parokonny AS, Gleba YY, Bennett MD (1993) Characterization of the *Nicotiana tabacum* L. genome by molecular cytogenetics. Mol Gen Genet 240:159–169

Kron P, Suda J, Husband BC (2007) Applications of flow cytometry to evolutionary and population biology. Ann Rev Eco Evol Syst 38:847–876

Leitch IL, Bennett MD (1997) Polyploidy in angiosperms. Trends Plant Sci 2:470–476

Leuchtenberger C (1958) Quantitative determination of DNA in cells by Feulgen microspectophotometry. In: Danielli JF (ed) General cytochemical methods. Academic Press, New York, pp 219–278

Li J, Das K, Liu J, Fu G, Li Y, Tobias C, Wu R (2012) Statistical models for genetic mapping in polyploids: challenges and opportunities. Methods Mol Biol 871:245–261

Liu B, Vega JM, Feldman M (1998) Rapid genomic changes in newly synthesized amphiploids of *Triticum* and *Aegilops*. II. Changes in low-copy coding DNA sequences. Genome 41:535–542

Ma JF, Shen R, Zhao Z, Wissuwa M, Takeuchi Y, Ebitani T, Yano M (2002) Response of rice to Al stress and identification of quantitative trait loci for Al tolerance. Plant Cell Physiol 43:652–659

Madlung A, Tyagi AP, Watson B, Jiang H, Kagochi T, Doerge RW, Martienssen R, Comai L (2005) Genomic changes in synthetic *Arabidopsis* polyploids. Plant J 41:221–230

McLeish J, Sunderland N (1961) Measurements of deoxyribonucleic acid (DNA) in higher plants by Feulgen photometry and chemical methods. Exp Cell Res 24:527–540

Nicolas SD, Leflon M, Monod H, Eber F, Coriton O, Huteau V, Chèvre AM, Jenczewski E (2009) Genetic regulation of meiotic cross-overs between related genomes in *Brassica napus* haploids and hybrids. Plant Cell 21:373–385

Oliver R, Jellen E, Ladizinsky G, Korol A, Kilian A, Beard J, Dumlupinar Z, Wisniewski-Morehead N, Svedin E, Coon M et al (2011) New Diversity Arrays Technology (DArT) markers for tetraploid oat (*Avena magna* Murphy et Terrell) provide the first complete oat linkage map and markers linked to domestication genes from hexaploid *A. sativa* L. Theor Appl Genet 123:1159–1171

Otto SP (2007) The evolutionary consequences of polyploidy. Cell 131:452–462

Paterson AH, Bowers JE, Chapman BA (2004) Ancient polyploidization predating divergence of the cereals, and its consequences for comparative genomics. Proc Nat Acad Sci U S A 101:9903–9908

Pires JC, Lim KY, Kovarik A, Matyasek R, Boyd A, Leitch AR, Leitch IJ, Bennett MD, Soltis PS, Soltis DE (2004) Molecular cytogenetic analysis of the recently evolved Tragopogon (Asteraceae) allopolyploids reveals a karyotype that is additive of the diploid progenitors. Am J Bot 91:1022–1035

Porceddu A, Albertini E, Barcaccia G, Falistocco E, Falcinelli M (2002) Linkage mapping in apomictic and sexual Kentucky bluegrass (*Poa pratensis* L.) genotypes using a two way pseudotestcross strategy based on AFLP and SAMPl markers. Theor Appl Genet 104:273–280

Raina SN, Raini V (2001) GISH technology in plant genome research. Methods Cell Sci 23:83–104

Reinisch AJ, Dong JM, Brubaker CL, Stelly DM, Wendel JF, Paterson AH (1994) A detailed RFLP map of cotton, *Gossypium hirsutum* x *Gossypium bar badense*: chromosome organization and evolution in a disomic polyploid genome. Genetics 138(3):829–847

Salmon A, Ainouche ML, Wendel JF (2005) Genetic and epigenetic consequences of recent hybridization and polyploidy in *Spartina* (Poaceae). Mol Ecol 14:1163–1175

Sampedro J, Yi L, Robert EC, de Pamphilis C, Cosgrove DJ (2005) Use of genomic history to improve phylogeny and understanding of births and deaths in a gene family. Plant J 44:409–419

Schlueter JA, Dixon P, Granger C, Grant D, Clark L, Doyle JJ, Shoemaker RC (2004) Mining EST databases to resolve evolutionary events in major crop species. Genome 47:868–876

Shi T, Huang H, Barker MS (2010) Ancient genome duplications during the evolution of kiwifruit (*Actinidia*) and related Ericales. Ann Bot 106:497–504

Shoemaker RC, PolzinK LJ, SpechtJ BEC, OlsonT YN, ConcibidoV WJ, Tamulonis JP, Kochert G, Boerma HR (1996) Genome duplication in soybean (*Glycine* subgenus *soja*). Genetics 144:329–338
Song K, Lu P, Tang K, Osborn TC (1995) Rapid genome change in synthetic polyploids of *Brassica* and its implications for polyploid evolution. Proc Natl Acad Sci USA 92:7719–7723
Stace CA (2000) Cytology and cytogenetics as a fundamental taxonomic resource for the 20th and 21st centuries. Taxon 49:451–477
Stebbins GL (1950) Variation and evolution in plants. Columbia University Press, New York
Sterck L, Rombauts S, Jansson S, Sterky F, Rouzé P, Van de Peer Y (2005) EST data suggest that poplar is an ancient polyploid. New Phytol 167(1):165–170
Szadkowski E, Eber F, Huteau V, Lode M, Huneau C, Belcram H, Coriton O, Manzanares-Dauleux MJ, Delourme R, King GJ et al (2010) The first meiosis of resynthesized *Brassica napus*, a genome blender. New Phytol 186:102–112
Udall JA, Wendel JF (2006) Polyploidy and crop improvement. Crop Sci 46:3–14
Van Dijk T, Noordijk Y, Dubos T, Bink M, Meulenbroek B, Visser R, van de Weg E (2012) Microsatellite allele dose and configuration establishment (MADCE): an integrated approach for genetic studies in allopolyploids. BMC Plant Biol 12. https://doi.org/10.1186/1471-2229-12-25
Vision TJ, Brown DG, Tanksley SD (2000) The origins of genomic duplications in *Arabidopsis*. Science 290:2114–2117
Wu KK, Burnquist W, Sorrells ME, Tew TL, Moore PH, Tanksley SD (1992) The detection and estimation of linkage in polyploids using single-dose restriction fragments. Theor Appl Genet 83:294–300
Wu R, Ma CX, Casella G (2004) A bivalent polyploid model for mapping quantitative trait loci in outcrossing tetraploids. Genetics 166:581–595
Yu J, Wang J, Lin W, Li S, Li H et al (2005) The genomes of *Oryza sativa*: a history of duplications. PLoS Biol 3:e38

Significance of Polyploidy

5

5.1 Advantages of Polyploidy

Polyploidy has played a significant role in the evolution of higher plants (Leitch and Bennett 1997), having helped in tremendously increasing the number of species and sometimes of genera on the planet Earth. It is the most rapid method known of producing radically different but vigorous and well-adapted genotypes (Stebbins 1950); for this reason, it is of great value in plant breeding, more importantly the allopolyploidy. The significance of polyploidy in the plant world can be visualized by considering its high incidence among angiosperms, including the leading commercial crops, such as wheat, sugarcane, oats, cotton, tobacco, apples, pears, plums, *Rubus*, etc. (Table 5.1). This incidence confers some important advantages, viz. heterosis, gene redundancy and asexual reproduction (Comai 2005).

5.1.1 Heterosis

Heterosis or hybrid vigour, in which the polyploid offspring of two diploid progenitors is healthier than either of the two parents, occurs in different ways. One is the enforced pairing of homologous chromosomes within an allotetraploid that prevents the recombination between the genomes of the original progenitors, thus effectively maintaining heterozygosity throughout generations. This heterozygosity prevents the accumulation of recessive mutations in the genomes of later generations, thereby maintaining hybrid vigour (Comai 2005). Heterosis is a permanent feature of allopolyploids; it is generally high in allopolyploids and is absent in autopolyploids. Dominant or overdominant gene actions are predicted to be the genetic basis of heterosis (reviewed in Chen 2010). Osborn et al. (2003) found that inter-genomic heterozygosity has positive effects on oilseed production in *Brassica napus*; the oilseed yield was less when this heterozygosity was lost. In nature, there exist several biological phenomena and one of the most important among these is 'heterosis', and it refers to superiority (phenotypic/agronomic) of heterozygous F_1 hybrids

T.-Ul.-H. Dar, R.-Ul. Rehman, *Polyploidy: Recent Trends and Future Perspectives*, https://doi.org/10.1007/978-81-322-3772-3_5

Table 5.1 Economically important polyploids

Taxa	Natural/synthetic	Ploidy auto/allo
Wheat (*Triticum aestivum*)	Synthetic	6X: Allopolyploid
Maize (*Zea mays*)	Synthetic	4X: Autopolyploid
Oats (*Avena sativa*)	Natural	6X: Allopolyploid
Alfalfa (*Medicago sativa*)	Natural	4X: Autopolyploid
Sugarcane (*Saccharum officinale*)	Natural	8X: Autopolyploid
Potato (*Solanum tuberosum*)	Natural	4X: Autopolyploid
Sweet potato (*Ipomoea batatas*)	Natural	6X:Autopolyploid
Banana (*Musa sapientum*)	Natural	2X, 3X: Autopolyploid
Cotton (*Gossypium hirsutum*)	Natural	4X: Allopolyploid
Tobacco (*Nicotiana tabacum*)	Natural	4X: Allopolyploid
Coffee (*Coffea arabica*)	Natural	4X: Allopolyploid
Apple (*Malus domestica*)	Natural	2X, 3X: Allopolyploid
Pear (*Pyrus communis*)	Natural	2X, 3X: Allopolyploid
Strawberry (*Fragaria cascadensis*)	Synthetic	10X: Allopolyploid
Peanut (*Arachis hypogaea*)	Natural	4X: Allopolyploid
Watermelon (*Citrullus lanatus*)	Synthetic	3X: Autopolyploid
Plum (*Prunus domestica*)	Natural	6X: Allotetraploid
Rapeseed (*Brassica napus*)	Synthetic	4X: Allotetraploid
Soybean (*Glycine max*)	Natural	4X: Allopolyploid
Lime (*Citrus aurantifolia*)	Natural	3X: Allopolyploid

when compared to genetically diverse (homozygous) parents. The superiority is demonstrated by hybrids over its parents with respect to several traits such as growth rate, biomass, yield parameters, disease resistance (microbes, insects) and climatic stresses (Shull 1952; Falconer and Mackay 1996). The heterosis phenomenon has been successfully applied in breeding programmes and in agronomic studies of many crops and animals for better production. However, hybrid plants demonstrate reduction in heterozygosity and vigour gradually over several generations due to self-pollination, and this phenomenon is referred to as inbreeding depression. Thus, heterosis and inbreeding depression are different aspects of the same phenomenon (Zirkle, 1952). The heterosis phenomenon was discovered by Charles Darwin in 1876 in *Zea mays* where he observed that cross-pollinated plants were 25% taller than the self-pollinated (inbred) plants (Darwin 1876). Subsequently, this phenomenon was rediscovered independently by George H. Shull and Edward M. East in 1908 (Shull 1908; East 1908). In *Zea mays*, heterosis is most evident in traits of later stages of development such as the cob size, but it is also manifested in the early stages of embryo and seedling development (Swanson-Wagner et al. 2006). The degree of phenotypic difference of a particular trait in a hybrid (F_1) in comparison to the parental inbred lines (P_1, P_2) is described by the terms MPH (mid-parent heterosis) and BPH (best parent heterosis). MPH is an indicative of the trait which displays significantly better performance in a hybrid than the average value of two parental inbred lines (MPH = F1 _ [(P1 + P2)/2]). On the other hand, BPH is indicative of the trait which displays significantly better performance in a hybrid than the better (Pb) of two homozygous parental inbred lines (BPH = F1 _ Pb).

5.1.1.1 Classical Genetic Explanation

The phenomenon of heterosis was rediscovered a century ago and many models have been suggested, but there is little consensus about its genetic basis (Lamkey and Edwards 1998; Stuber 1999; Birchler et al. 2003, 2006).The explanations for heterosis in classical genetics centres on three hypotheses, i.e. dominance (Davenport 1908; Bruce 1910; Keeble and Pellow 1910; Jones 1917), overdominance (East 1908; Shul 1908) and epistasis (Powers 1944; Yu et al. 1997). Among these hypotheses, the two most prominent are 'dominance' and 'overdominance'. Both of these hypotheses ascribe the superior phenotypic behaviour as a consequence of genetic difference between parental inbred lines (homozygous) and their hybrids (heterozygous). The term dominance refers to the vigour of hybrid plants which is due to complimentary action of the dominant alleles from the two parental inbred lines at multiple loci over the corresponding unfavourable alleles (Davenport 1908; Keeble and Pellow 1910; Bruce 1910; Jones 1917), while overdominance refers to the superior hybrid traits in comparison to the parental inbred lines to allelic interactions at one or multiple loci (Shull 1908; Hochholdinger and Hoecker 2007). Finally, the third hypothesis is epistasis (Powers 1945) which refers to superiority of phenotypic expression of a trait in the hybrid due to interaction between non-allelic genes at two or more loci. However, heterosis has also been shown in tomato introgression lines in the absence of epistasis (Semel et al. 2006). Based on these hypotheses, it can be summarized that the vigorous phenotypes in hybrids over the inbred lines are due to various combinations of several genes which implies that positive and negative effects of the various loci might be compensating each other and thus making it difficult to choose one hypothesis over the other (Melchinger 1999). Further, it must be noted that the quantitative genetics terms, viz. dominance and overdominance, should not be associated directly with the quantitative behaviour of the phenotypic traits or the molecular principles (Hochholdinger and Hoecker 2007).

5.1.1.2 Molecular Basis of Heterosis

The above classical genetics hypotheses of heterosis are said to be largely conceptual and are not connected to molecular principles and are thus insufficient to explain the molecular basis of heterosis (Birchler et al. 2003). There are many studies in the recent past which have explored the molecular mechanisms of heterosis at the genome-wide level in plants. These studies suggest that the heterosis in plants is contributed potentially by the global differential gene expression between the hybrids and parental lines (Swanson-Wagner et al. 2006; Zhang et al. 2008). Among the genetic components, the cis-acting elements and the trans-acting factors are reported to be the key regulators of differential gene expression in hybrids (Stupar and Springer 2006; Zhang et al. 2008). Further, the other potential impactors of heterosis are the epigenetic components, regulators of chromatin states and genome activity (Ha et al. 2009; He et al. 2010; Groszmann et al. 2011; Barber et al. 2012; Chodavarapu et al. 2012; Shen et al. 2012). The main eukaryotic epigenetic components are DNA cytosine methylation, histone modifications and certain aspects of RNA interference (RNAi) (Henderson and Jacobsen 2007; Zhang et al. 2008). There is a possibility of propagating naturally occurring variations in the above epigenetic components across the generations which can lead to significant heritable variation in phenotypic traits (Richards 2006; Richards et al. 2010). Heritable epigenetic variations usually coexist with sequence-based genetic variations at the same genomic loci, making it difficult to discern between the effects of them on the determination of phenotype. According to Richards, the range of epigenetic variation in relation to genotypic context could be classified into three categories: obligatory, facilitated and pure. Whereas obligate epigenetic variation is completely dependent on genetic variation, facilitated epigenetic variation is semi-independent of genetic variation, and pure epigenetic variation is completely independent of genetic variation (Richards 2006). Recently, many studies have surveyed the global epigenetic variations (i.e. epigenomic variation) and their relationships with genetic variations and phenotypic diversity in different plant species (He et al. 2011). Of particular interest, some studies have explored simultaneously the epigenomic and transcriptomic variation among hybrids and their parental lines (He et al. 2010; Shen et al. 2012). These pilot studies provide us an opportunity to understand the molecular mechanisms of heterosis in plants from the view of epigenetics. In this review, we will briefly summarize recent studies on epigenetic variations in hybrids and their potential associations with heterosis in plants.

5.1.1.3 Role of DNA Methylation in Heterosis

In eukaryotic system, the DNA methylation takes place by the addition of a methyl group by DNA methyltransferases at the 5 position of cytosine base. The DNA methylation in plants takes place in CG and CHG (H is ¼ A, T or C) and asymmetric (CHH) contexts catalysed by DNA methyltransferase 1(MET1), chromomethyltransferase 3 (CMT3) and domain rearranged methyltransferase 2 (DRM2) (Chan et al. 2006; Law and Jacobsen 2010). In plants removal of methylated cytosines, i.e. demethylation, can be done by enzymes DNA glycosylases DEMETER (DME), REPRESSOR OF SILENCING 1 (ROS1), and DEMETER-LIKE 2 (DML2) and DML3. These enzymes have been reported to be involved in several important processes such as in development and physiology (Zhu 2009). In plants the process of DNA methylation takes place particularly on transposons and other repetitive elements. DNA methylation is thus regarded as the mechanism of repressive epigenetics which maintains the genomic stability by suppressing the activity of above elements (Law and Jacobsen 2010). DNA methylation in plants has also been shown by genome-wide studies to be associated with the promoter or transcribed regions of protein-coding genes, thus suggesting its role in regulating gene activity (Zhang et al. 2006; Zilberman et al. 2007; Li et al. 2008; He et al. 2010). By employing the methylation-sensitive amplified polymorphism (MSAP) method, the alterations in cytosine methylation in several hybrid plants have been observed such as maize (Zhao et al. 2007), rice (Jin et al. 2008), *Arabidopsis* (Banaei Moghaddam et al. 2010) and Phlox (Dar et al. 2013) intraspecific hybrids. Furthermore, another method of combining McrBC digestion with deep sequencing has helped to generate genome-wide

DNA methylation profiles in two rice subspecies and their reciprocal hybrids leading to the identification of several genes and their variation in terms of DNA methylation, e.g. transposable element-related genes (He et al. 2010). In another study by utilizing high-throughput bisulphite sequencing (MethylC-seq), DNA methylomes were generated in two *Arabidopsis* ecotypes and their reciprocal hybrids at single base-pair resolution. The methylomes were altered in hybrids as such that CG methylation increased and CHH methylation decreased in comparison to parents, and it was observed that the changes in DNA methylation occurred frequently at the loci where parental methylation levels were different (Greaves et al. 2012a). In another study using the above *Arabidopsis* hybrids, it was further revealed that the cytosine methylation increased in all sequence contexts across the entire genome (especially in transposable elements), and it was found that increase in DNA methylation in hybrids was due to the differences in parental DNA methylation (Shen et al. 2012), and these results were also corroborated in another study with rice (Chodavarapu et al. 2012). Further, it has been argued that the reduction in DNA methylation could affect the hybrid vigour thus providing the direct correlation between the DNA methylation and its role in heterosis (Shen et al. 2012).

5.1.1.4 Role of Small RNAs in Heterosis

There are two major types of small RNAs in plants, viz. microRNAs (miRNAs) and small interfering RNAs (siRNAs) (Chen 2009). Primarily, the role of miRNAs is in regulation of target mRNAs by the transcript cleavage and translational inhibition, while siRNAs mostly exist as heterochromatic siRNA or repeat-associated siRNA (rasiRNA). The Heterochromatic siRNAs are produced from transposable elements and other repeat sequences, and they maintain the genome stability by silencing transposable elements through RNA-directed DNA methylation (RdDM) pathway (Henderson and Jacobsen 2007; Chen 2009). It has been shown in many studies that there is an extensive variation in siRNA transcription in *Arabidopsis*, rice, maize and wheat hybrids (Ha et al. 2009; Groszmann et al. 2011; Li et al. 2012; Shen et al. 2012; Chen et al. 2010; He et al. 2010; Chodavarapu et al. 2012; Barber et al. 2012; Ding et al. 2012; Kenan-Eichler et al. 2011). There was an additive expression pattern of most miRNAs or siRNA loci in hybrids. The nonadditive expression of miRNAs of these hybrids was associated with differential expression of protein-coding genes when compared to parental lines (Ha et al. 2009; Chen et al. 2010; He et al. 2010; Ding et al. 2012). It has been observed in hybrids that siRNAs show nonadditive variation pattern such as downregulation trend after hybridization (He et al. 2010; Groszmann et al. 2011; Barber et al. 2012; Li et al. 2012; Shen et al. 2012). Further it was also found that in the genomic region, the siRNAs were associated with the genes and their flanking regions, thus suggesting the role of siRNAs in variation of gene expression in hybrids (Groszmann et al. 2011). Upon analysis, it was found that variations of DNA methylation in hybrids were associated with siRNA expression which suggested that the transchromosomal methylation and transchromosomal demethylation are mediated by RdDM pathway. The siRNA expression could lead to either hyper- or hypomethylation of two parental alleles (Groszmann et al. 2011; Chodavarapu et al. 2012; Greaves et al. 2012; Shen et al.

2012). However, there are few recent studies with the mutants which have reached to different conclusion with respect to the role of siRNAs in heterosis. In case of Hua Enhancer 1 (HEN 1: encodes RNA methyltransferase crucial for plant RNAs biogenesis) *Arabidopsis* mutants, the heterosis declined in hen1 hybrids (Shen et al. 2012). In MOP 1 (MODIFIER OF PARAMUTATION1: heterochromatic siRNA biogenesis) maize mutants, 24-nt siRNAs were reduced without any effect on heterosis (Barber et al. 2012). This opposite effect between the hen1 and mop1 mutants could be due to difference in RNA species (Nobuta et al. 2008). These conflicting evidences on the heterosis need to be further ascertained in the future.

5.1.1.5 Role of Histone Modification in Heterosis

The core histone proteins (which constitute nucleosome) undergo various post-translational covalent modifications (PTMs) such as methylation, acetylation, phosphorylation, ubiquitylation, glycosylation, ADP-ribosylation, sumoylation, deimination and proline isomerization (Berger 2007; Kouzarides 2007). These histone modifications are responsible for regulating the chromatin state and genomic activity during various developmental and environmental cues. The most widely studied PTMs are histone acetylation and methylation at lysine/residues. The histone lysine acetylation is done by histone acetyltransferases (HATs) and the deacetylation is done by histone deacetylases (HDACs), and these are associated with gene activation, while the histone lysine methylation is helped by histone lysine methyltransferases (HKMTs) and demethylation is done by histone demethylases (HDMs), and these may be associated with transcriptional activation or repression (Berger 2007; Kouzarides 2007; Li et al. 2007; Shahbazian and Grunstein 2007; Liu et al. 2010; He et al. 2011). For the first time, the association between histone modification and the heterosis process was reported by Ni et al. (2009), demonstrating that the alteration of circadian rhythms in *Arabidopsis* hybrids was due to the association between the circadian clock genes CCA1 (Circadian Clock Associated 1) and LHY (Late Elongated Hypocotyl) and these genes contributed to growth vigour and biomass increase in these hybrids. Upon examination, it was revealed that the transcriptional repression of these circadian clock genes (CCA1 and LHY) in hybrids was due to lowering of two histone modifications in the promoter regions of these genes. In other study employing the rice hybrids with either overexpressed or inactivated histone deacetylase gene OsHDT1 resulted in the affect on nonadditive expression on several genes (Li et al. 2011). The survey on the genome-wide variation in rice hybrids and their parents in two active histone marks (H3K4me3 and H3K9ac) and a repressive histone mark (H3K27me3) has revealed that the inheritance of histone modifications was relatively stable and additive in hybrid offsprings (He et al. 2010). This observation was also made in *Arabidopsis* (Banaei Moghaddam et al. 2011). Interesting in rice, the nonadditive inheritance of histone modifications in hybrids is ascribed to the changes in the transcript level between hybrids and parents. The allelic bias was exhibited in reciprocal hybrids due to histone modifications suggesting that there is a differential contribution of parental alleles to the epigenetic states of genes in hybrids (He et al. 2010) (Fig. 5.1).

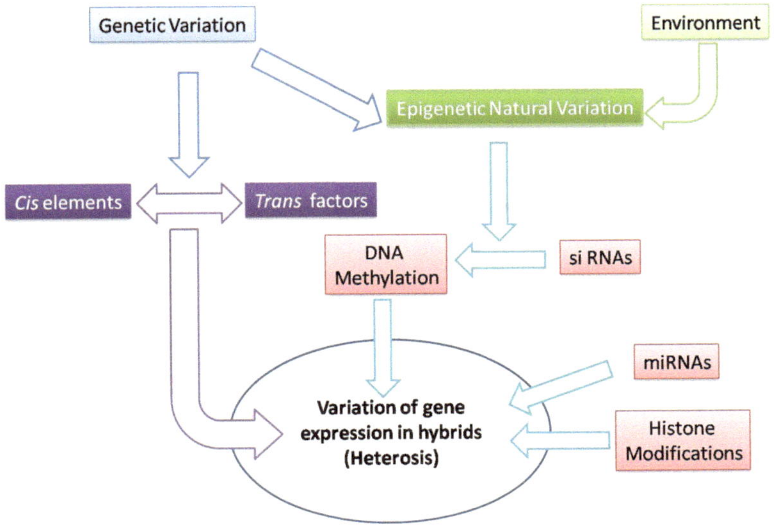

Fig. 5.1 Heterosis: regulation of gene expression

5.2 Gene Redundancy

As polyploids have more than two sets of genes, they can withstand the deleterious effects of recessive mutations. This is particularly important during the gametophyte life stage. During the haploid phase of the life cycle, any allele that is recessive for a deleterious mutation will not be masked by the presence of a dormant, normally functioning allele, allowing the mutation to cause developmental failure in the pollen or egg sac. Conversely, a diploid gamete permits the masking of this deleterious allele by the presence of dominant normal allele, thereby protecting the pollen or egg sac from the developmental dysfunction (Comai 2005). Another advantage conferred by gene redundancy is the ability to diversify gene function over time. In other words, extra copies of genes that are not required for normal function might end up being used in new and different ways, leading to new opportunities in evolutionary selection (Adams and Wendel 2005).

5.3 Promoting Self-Fertilization and Asexual Reproduction

Polyploidy can disrupt certain self-incompatible systems and thereby allow self-fertilization (Miller and Venable 2000). This might be the result of the interactions between parental genomes in allopolyploids (Comai et al. 2000; Comai 2005). The ability to self-fertilize reduces the effect of minority cytotype disadvantage by eliminating the need for a cytoplasmically compatible mate. For this reason, it may be that the selfing taxa, on an average, produce successfully more polyploids than do outcrossing taxa (Barringer 2007).

5.4 Disadvantages

Since naturally occurring polyploidy seems to be most often associated with hybridization, its effects in nature may be considered hazardous (Stebbins 1950). To ascertain this, many workers have studied the effects of artificially induced polyploidy on various species in nature (reviewed in Noggle 1946). Based on such studies, it is now believed that genome doubling may sometimes be complicated and may influence the functioning of an individual. Following are some of the effects of polyploidy:

5.4.1 Cellular Changes

Increase in size of individual cells is the primary effect of polyploidy. Upon polyploidy, the genome doubles, increasing volume of the cell. This is because the cell volume is proportional to the amount of DNA in the cell nucleus. Doubling a cell's genome is expected to double the volume of space occupied by the chromosomes in the nucleus, but actually it causes only 1.6-fold increase in the surface area of the nuclear envelope (Melaragno et al. 1993). This can disrupt the balance of factors that normally mediate interaction between the chromosomes and nuclear components, including envelop-bound proteins (Comai 2005). Altered geometric relations between the nucleus and the rest of the cell also result in a number of secondary effects, viz. increase in water content of cells; differences in contents of proteins, chlorophyll, cellulose, auxin and vitamins; changes in growth rate (metabolism and growth seem to be rather retarded in polyploids); late flowering; and reduced branching.

5.4.2 Mitotic and Meiotic Irregularities

Polyploidy can also be problematic for the normal completion of mitosis and meiosis. Aneuploid cells, which have abnormal number of chromosomes, are more readily produced in meiosis involving three or more sets of chromosomes than in diploid cells. Autopolyploids have the potential to form multiple arrangements of homologous chromosomes at meiotic metaphase I, which can result in abnormal segregation patterns, such as 3:1 or 2:1 plus one laggard. These abnormal segregation patterns cannot be resolved into balanced products, and random segregation of multiple chromosome types produces mostly aneuploid gametes (Comai 2005).

5.4.3 Deviated Gene Activity

An increase in the copy number of all chromosomes affects all genes equally and should result in a uniform increase in gene expression. However, it is possible that some genes deviate from this assumption because they respond to regulating factors that do not change proportionally with ploidy (Comai 2005). For example, Guo et al. (1996) compared the mRNA levels per genome for 18 genes in 1x, 2x, 3x and 4x maize. It was observed that although expression of most genes increased with ploidy, some other genes showed deviated expression. For example, sucrose synthase showed the expected proportional expression in 2x and 4x tissues, but its expression was three and six times higher, respectively, in 1x and 3x tissues. Two other genes showed similar, but less extreme, trends. Altogether, about 10% of these genes demonstrated sensitivity to odd-number ploidy (Comai 2005).

5.4.4 Lowering of Growth Hormones

So far as growth hormones are concerned, lower content of abscisic acid and auxin-like substances was found in the autotetraploid *Lycopersicon pimpinellifolium*, as compared to diploids (Lewis 1980).

5.4.5 Effect on Fertility and Genetic Behaviour

The most conspicuous and significant effect of polyploidy is the reduction of pollen and seed fertility in autopolyploids as compared to their diploid ancestors (Stebbins 1950).

5.5 Phenotypic Changes of Polyploidy

Polyploidization has direct effect on the phenotype. Chromosome doubling brings some abrupt, transgressive and conspicuous changes in populations within micro-evolutionary time (Levin 1983). Identifying natural polyploids that differ only in ploidy is sometimes difficult. But due to the presence of traits like hybridity and asexual reproduction which frequently occur in polyploid lineages can be considered the effects of polyploidy. In allopolyploids, it is difficult to separate the effects of polyploidy from the effects of hybridity.

5.5.1 Effects on Cell Size and Growth Rates

The most common effect of genome doubling is increased cell volume. It may depend upon environmental conditions. Usually stomatal cell size is initially used as an indirect measure of ploidy level. Pollen grain diameter is also used to detect ploidy. There are some reports from some species of amphibian (Beatty 1957) and plants (Butterfass 1987) that by the passage of time, the cell volume of polyploids tends to decrease as related to diploids. Changes in cell volume may alter the metabolic process especially those in membranes (Cavalier-Smith 1978). These metabolic changes are likely associated with growth rates and detected in polyploids. Polyploids generally have slower developmental rates than related diploids. Seed size is also large in polyploids than in diploids.

5.5.2 Effects on Size and Shape

Polyploidy is sometimes but not always associated with larger overall size (Stebbins 1950). Increased size has been observed in invertebrates (Bell 1982), but in vertebrates, little effect has been observed on body size (Mahony and Robinson 1980; Schultz 1980). Broader and thicker leaves, fewer stems per plant and fewer inflorescences, increased seed weight, but fewer seeds overall had been often associated with polyploids.

5.5.3 Effects on the Reproductive Systems

Shifts in the reproductive system towards asexual reproduction correlate with the occurrence of polyploidy. Most apomictic plants are polyploid, while most are not apomictic. Majority of polyploid fish and amphibian reproduce sexually, but most of polyploid insects and reptiles reproduce parthenogenetically. In animals, parthenogenesis arises in diploids, and these diploids mate with related sexes and generate polyploids. These polyploids are generated through the union of unreduced eggs and sperms. In flowering plants, polyploidization often appears to predate apomixes (Stebbins 1980).

In addition to asexual reproduction, other breeding systems have been noted in plants. Self-incompatibility breaks down in polyploids, resulting in higher selfing rates in polyploids than in their diploid progenitors (Stebbins 1950). Polyploidization changes floral traits, including relative sizes and spatial relation of floral organs (Brochmann et al. 1993).

5.5.4 Geographical Distribution and Ecological Correlates

Polyploids have wider ecological distribution than their diploid progenitors (Levin 1983; Sexton 1980). Increased heterozygosity has been found to be responsible for providing metabolic flexibility which in turn can cope with broader array of conditions. For example, a particular polyploid has different forms of enzyme available due to heterozygosity. They may perform function in different conditions. Polyploids can exist in more than one niche than their diploid progenitors. Niche portioning is favourable for newly established polyploids (Martins et al. 1998).

Polyploids frequently occur at higher altitudes and more polar latitudes than diploids (Barata et al. 1996; Beaton and Hebert 1988; Bell 1982). In polar regions, cold treatments can increase the frequency of unreduced gamete formation, and these are involved in polyploidization events (Bogart et al. 1989; Fankhauser 1945; Ramsey and Schemske 1998). Larger cell size may be responsible for the selective advantage of some polyploids in these areas (Dufresne and Hebert 1998).

5.6 Survival in Adverse Environments

The frequency of polyploidy is more at high altitudes and latitudes and might be resistant to dryer conditions revealing their fitness advantage (Grant 1981; Maherali et al. 2009; Brochmann et al. 2004). According to evolutionary studies in angiosperms, the timing of polyploidization coincides with timing of Cretaceous-Tertiary boundary that has led to the hypothesis that genome doubling helped in better adaptation under severe environment, than their diploid relative (Fawcett et al. 2009). However, while considering environmental effects on gamete formation, it might be possible that the climatic changes during mass extinction events resulted in unreduced gamete formation leading to formation of polyploids at much faster pace.

It has also been found that polyploids succeeded more in colonizing arctic regions after the deglaciation than the diploids (Brochmann et al. 2004). A worldwide study on endangered species (640) and on invasive species (81) found that endangered species were mostly diploid, while invasive species were polyploids suggesting that polyploidization might be providing tolerance in diverse extreme environments (Pandit et al. 2011). In *A. thaliana*, polyploids were seen to be more tolerant having better reproductive success than diploid cytotypes under salinity stress (Chao et al. 2013). However, all findings do not support the notion of better survival of polyploids as reported in a study on 100 polyploid and diploid species where no correlation was found between polyploidization and winter hardiness (Bowden 1940). Similarly in another study in North America, no association was found between ploidy level, species range area, temperature, precipitation and latitudes (Martin and Husband 2009). Thus it can be conferred from the above studies that physiological changes happening due to ploidy do not necessarily provide superior abiotic and biotic stress tolerance. This suggests that relationship with adaptation is complex and it does not depend only on species undergoing polyploidization but also on its growing environment.

5.7 Reproduction and Isolating Mechanisms

For the polyploidy species to be successful, the newly formed polyploids should be able to establish and show persistence to the new environment. These processes have been studied by Stebbins (1950) in the light of isolating mechanisms. There are a number of features which contribute to the reproductive success of the species such as perenniality (propensity towards apomixes or self-compatibility) and morphological features (floral characteristics) (e.g. Tate et al. 2005). The newly formed polyploids might be reproductively isolated from their progenitors due to multiple chromosome numbers resulting in the alteration of reproductive biology (physiological and morphological changes). The reproductive barriers between polyploids and diploid progenitors may be prezygotic (pollinator consistency, phenology differences and geographical isolation) and post-zygotic (e.g. inbreeding depression and triploid hybrid inviability). The genetic incompatibility system breakdown (Richards 1997) may result in polyploidy formation leading to self-fertilization which can result in isolating the new polyploid from its diploid progenitors and thus promotes polyploidy establishment. This has been reported to occur in species single-locus gametophytic self-incompatibility system (de Nettancourt 2001; Ramsey and Schemske 2002). The autotetraploids of *Chamerion angustifolium* showed self-pollination, pollinator fidelity, flower asynchrony and gametic selection (Husband and Sabara 2003).

While emphasizing on post-zygotic factors such as triploid sterility/invariability, it has been proposed that it might be secondary to prezygotic isolation mechanisms giving enough time for evolution for prezygotic barriers. The polyploidization affecting morphological changes might have an impact on the reproductive biology of new polyploidy species, e.g. it was shown that autopolyploid populations of *Heuchera grossulariifolia* had larger flowers, variable flower shapes and different timing of flowering than diploid populations. Further, it was also seen that pollinators visiting diploids and tetraploids differed even though the flowering time of the two was in synchrony (Segraves and Thompson 1999). Furthermore, it has been also seen that the independently generated autotetraploids too differed in floral characteristics.

References

Adams KL, Wendel JF (2005) Allele-specific, bidirectional silencing of an alcohol dehydrogenase gene in different organs of interspecific diploid cotton hybrids. Genetics 171:2139–2142

Banaei Moghaddam AM, Fuchs J, Czauderna T, Houben A, Mette MF (2010) Intraspecific hybrids of Arabidopsis thaliana revealed no gross alterations in endopolyploidy, DNA methylation, histone modifications and transcript levels. Theor Appl Genet 120(2):215–226

Banaei Moghaddam AM, Roudier F, Seifert M, Berard C, Magniette MLM, Ashtiyani RK, Houben A, Colot V, Mette MF (2011) Additive inheritance of histone modifications in Arabidopsis thaliana intra-specific hybrids. Plant J 67:691–700

Barata C, Hontoria F, Amat F, Browne R (1996) Competition between sexual and parthenogenetic Artemia: temperature and strain effects. J Exp Mar Biol Ecol 196:329–340

Barber WT, Zhang W, Win H, Varala KK, Dorweiler JE, Hudson ME, Moose SP (2012) Repeat associated small RNAs vary among parents and following hybridization in maize. Proc Natl Acad Sci U S A 109:10444–10449

Barringer BC (2007) Polyploidy and self-fertilization in flowering plants. Am J Bot 94:1527–1533

Beaton MJ, Hebert PDN (1988) Geographic parthenogenesis and polyploidy in *Daphnia pulex* Leydig. Am Nat 132:837–845

Beatty RA (1957) Parthenogenesis and polyploidy in mammalian development. Cambridge University Press, London/New York

Bell G (1982) The masterpiece of nature. University of California Press, Berkely

Berger SL (2007) The complex language of chromatin regulation during transcription. Nature 447:407–412

Birchler JA, Auger DL, Riddle NC (2003) In search of the molecular basis of heterosis. Plant Cell 15:2236–2239

Birchler JA et al (2006) Unraveling the genetic basis of hybrid vigor. Proc Natl Acad Sci U S A 103:12957–12958

Bogart JP, Elinson RP, Licht LE (1989) Temperature and sperm incorporation in polyploid salamanders. Science 246:1032–1034

Bowden WM (1940) Diploidy, polyploidy, and winter hardiness relationships in the flowering plants. Am J Bot 27:357–371

Brochmann C, Borgen L, Stedje B (1993) Crossing relationships and chromosome numbers of Nordic populations of Draba (Brassicaceae), with emphasis on the *Draba alpina* complex. Nord J Bot 13:121–147

Brochmann C, Brysting AK, Alsos IG, Borgen L et al (2004) Polyploidy in arctic plants. Biol J Linn Soc 82:521–536

Bruce AB (1910) The Mendelian theory of heredity and the augmentation of vigor. Science 32:627–628

Butterfass TH (1987) Cell volume ratios of natural and of induced tetraploid and diploid flowering plants. Cytologia 52:309–316

Cavalier-Smith T (1978) Nuclear volume control by nucleoskeletal DNA, selection for cell volume and cell growth rate, and the solution of the DNA C-value paradox. J Cell Sci 34:247–278

Chan S, Henderson I, Zhang X, Shah G, Chien J, Jacobsen S (2006) RNAi, DRD1, and histone methylation actively target developmentally important non-CG DNA methylation in arabidopsis. PLoS Genet 2:791–797

Chao DY, Dilkes B, Luo H, Douglas A et al (2013) Polyploids exhibit higher potassium uptake and salinity tolerance in Arabidopsis. Science 341:658–659

Chen X (2009) Small RNAs and their roles in plant development. Annu Rev Cell Dev Biol 25:21–44

Chen ZJ (2010) Molecular mechanisms of polyploidy and hybrid vigor. Trends Plant Sci 15:57–71

Chen F, He G, He H, Chen W, Zhu X, Liang M, Chen L, Deng XW (2010) Expression analysis of miRNAs and highly-expressed small RNAs in two rice subspecies and their reciprocal hybrids. J Integr Plant Biol 52:971–980

Chodavarapu RK, Feng S, Ding B, Simon SA, Lopez D, Jia Y, Wang GL, Meyers BC, Jacobsen SE, Pellegrini M (2012) Transcriptome and methylome interactions in rice hybrids. Proc Natl Acad Sci U S A 109:12040–12045

Comai L (2005) The advantages and disadvantages of being polyploid. Nat Rev Genet 6:836–846

Comai L, Tyagi AP, Winter K, Holmes Davis R, Reynolds SH, Stevens Y, Byers B (2000) Phenotypic instability and rapid gene silencing in newly formed Arabidopsis allotetraploids. Plant Cell 12:155–168

Dar TH, Raina SN, Goel S (2013) Molecular analysis of genomic changes in synthetic autotetraploids *Phlox drummondii* Hook. Biol J Linn Soc 110:591–605

Darwin CR (ed) (1876) The effects of cross- and self-fertilization in the vegetable kingdom, 1st edn. John Murray, London

Davenport CB (1908) Degeneration, albinism and inbreeding. Science 28:454–455

De Nettancourt D (2001) Incompatibility and incongruity in wild and cultivated plants, 2nd edn. Springer, Berlin

Ding D, Wang Y, Han M, Fu Z, Li W, Liu Z, Hu Y, Tang J (2012) MicroRNA transcriptomic analysis of heterosis during maize seed germination. PLoS One 7:e39578

Dufresne F, Hebert PDN (1998) Temperature-related differences in life history characteristics between diploid and polyploid clones of the Daphnia pulex complex. Ecoscience 5:433–437

East EM (1908) Inbreeding in corn. Conn Agric Exp Stat Rpt 1907:419–428

Falconer DS, Mackay TFC (eds) (1996) Introduction to quantitative genetics, 4th edn. Longman, Harlow

Fankhauser G (1945) The effects of changes in chromosome number on amphibian development. Q Rev Biol 20:20–78

Fawcett JA, Maere S, Van de Peer Y (2009) Plants with double genomes might have had a better chance to survive the cretaceous-tertiary extinction event. Proc Natl Acad Sci U S A 106:5737–5742

Grant V (1981) Plant speciation. Columbia University Press, New York

Greaves IK, Groszmann M, Ying H, Taylor JM, Peacock WJ, Dennis ES (2012) Trans chromosomal methylation in Arabidopsis hybrids. Proc Natl Acad Sci U S A 109:3570–3575

Groszmann M, Greaves IK, Albertyn ZI, Scofield GN, Peacock WJ, Dennis ES (2011) Changes in 24-nt siRNA levels in Arabidopsis hybrids suggest an epigenetic contribution to hybrid vigor. Proc Natl Acad Sci U S A 108:2617–2622

Guo M, Davis D, Bircher JA (1996) Dosage effects on gene expression in a maize ploidy series. Genetics 142:1349–1355

Ha M, Lu J, Tian L, Ramachandran V, Kasschau KD, Chapman EJ, Carrington JC, Chen X, Wang XJ, Chen ZJ (2009) Small RNAs serve as a genetic buffer against genomic shock in Arabidopsis interspecific hybrids and allopolyploids. Proc Natl Acad Sci U S A 106:17835–17840

He G, Zhu X, Elling AA, Chen L, Wang X, Guo L, Liang M, He H, Zhang H, Chen F, Qi Y, Chen R, Deng XW (2010) Global epigenetic and transcriptional trends among two rice subspecies and their reciprocal hybrids. Plant Cell 22:17–33

He G, Elling AA, Deng XW (2011) The epigenome and plant development. Annu Rev Plant Biol 62:411–435

Henderson IR, Jacobsen SE (2007) Epigenetic inheritance in plants. Nature 447:418–424

Hochholdinger F, Hoecker N (2007) Towards the Molecular Basis of Heterosis. Trends Plant Sci 12:427–432

Husband BC, Sabara HA (2003) Reproductive isolation between autotetraploids and their diploid progenitors in fi reweed, *Chamerion angustifolium* (Onagraceae). New Phytol 161:703–713

Jin J, Huang W, Gao JP, Yang J, Shi M, Zhu MZ, Luo D, Lin HX (2008) Genetic control of rice plant architecture under domestication. Nat Genet 40(11):1365–1369

Jones DF (1917) Dominance of linked factors as a means of accounting for heterosis. Genetics 2:466–479

Keeble F, Pellow C (1910) The mode of inheritance of stature and time of flowering in peas (*Pisum sativum*). J Genet 1:47–56

Kenan-Eichler M, Leshkowitz D, Tal L, Noor E, Melamed-Bessudo C, Feldman M, Levy AA (2011) Wheat hybridization and polyploidization results in deregulation of small RNAs. Genetics 188:263–272

Kouzarides T (2007) Chromatin modifications and their function. Cell 128:693–705

Lamkey KR, Edwards JW (1998) Heterosis: theory and estimation. In: Proceedings 34th Illinois corn breeders' school, Urbana, IL 2–3 March 1998, pp 62–77

Law JA, Jacobsen SE (2010) Establish patterns in plants and animals. Nat Rev Genet 11:204–220

Leitch IL, Bennett MD (1997) Polyploidy in angiosperms. Trends Plant Sci 2:470–476

Levin DA (1983) Polyploidy and novelty in flowering plants. Am Nat 122(1):1–25

Lewis WH (1980) Polyploidy in species populations. In: Lewis WH (ed) Polyploidy: biological relevance. Plenum, New York, pp 103–144

Li B, Carey M, Workman JL (2007) The role of chromatin during transcription. Cell 128:707–719

Li X, Wang X, He K, Ma Y, Su N, He H, Stolc V, Tongprasit W, Jin W, Jiang J et al (2008) High-resolution mapping of epigenetic modifications of the rice genome uncovers interplay between DNA methylation, histone methylation, and gene expression. Plant Cell 20:259–276

Li C, Huang L, Xu C, Zhao Y, Zhou DX (2011) Altered levels of histone deacetylase OsHDT1 affect differential gene expression patterns in hybrid rice. PLoS One 6:e21789

Li Y, Varala K, Moose SP, Hudson ME (2012) The inheritance pattern of 24 nt siRNA clusters in arabidopsis hybrids is influenced by proximity to transposable elements. PLoS One 7:e47043

Liu C, Lu F, Cui X, Cao X (2010) Histone methylation in higher plants. Annu Rev Plant Biol 61:395–420

Maherali H, Walden AE, Husband BC (2009) Genome duplication and the evolution of physiological responses to water stress. New Phytol 184:721–731

Mahony MJ, Robinson ES (1980) Polyploidy in the Australian leptodactylid frog genus Neobatrachus. Chromosoma 81:199–212

Martin SL, Husband BC (2009) Influence of phylogeny and 1 ploidy on species ranges of North American angiosperms. J Ecol 97:913–922

Martins MJ, Collares-Pereira MJ, Cowx IG, Coelho MM (1998) Diploids v. triploids of *Rutilus alburnoides*: spatial segregation and morphological differences. J Fish Biol 52:817–828

Melaragno JE, Mehrotra B, Coleman AW (1993) Relationship between endopolyploidy and cell size in epidermal tissue of Arabidopsis. Plant Cell 5:1661–1668

Melchinger AE (1999) Genetic diversity and heterosis. In: Coors JG, Pandey S (eds) The genetics and exploitation of Heterosis in crops. American Society of Agronomy, Inc/Crop Science Society of America, Inc, Madison, pp 99–118

Miller JS, Venable DL (2000) Polyploidy and the evolution of gender dimorphism in plants. Science 289(5488):2335–2338

Ni Z, Kim ED, Ha M, Lackey E, Liu J, Zhang Y, Sun Q, Chen ZJ (2009) Altered circadian rhythms regulate growth vigour in hybrids and allopolyploids. Nature 457:327–331

Nobuta K, Lu C, Shrivastava R, Pillay M, De Paoli E, Accerbi M, Arteaga-Vazquez M, Sidorenko L, Jeong DH, Yen Y, Green PJ, Chandler VL, Meyers BC (2008) Distinct size distribution of endogenous siRNAs in maize: evidence from deep sequencing in the mop1-1 mutant. Proc Natl Acad Sci U S A 105:14958–14963

Noggle GR (1946) The physiology of polyploidy in plants. I. Review of the literature. Lloydia 9:153–173

Osborn TC, Pires JC, Birchler JA, Auger DL, Chen ZJ, Lee HS, Comia L, Madlungh A, Doerge RW, Colot V, Matrienssen RA (2003) Understanding mechanisms of novel gene expression in polyploids. Trends Genet 19:141–147

Pandit MK, Pocock MJO, Kunin WE (2011) Ploidy influences rarity and invasiveness in plants. J Ecol 99:1108–1115

Powers L (1944) Meiotic studies of crosses between Fragaria ovales and F. ananassa. J Agr Res 69:435–448

Powers L (1945) An expansion of Jones's theory for the explanation of heterosis. Am Nat 78:275–280

Ramsey J, Schemske DW (1998) Pathways mechanisms and rates of polyploid formation in flowering plants. Annu Rev Ecol Syst 29:467–501

Ramsey J, Schemske DW (2002) Neopolyploidy in flowering plants. Annu Rev Ecol Syst 33:589–639

Richards AJ (1997) Plant breeding systems, 2nd edn. Springer/Chapman, & Hall, London

Richards EJ (2006) Inherited epigenetic variation-revisiting soft inheritance. Nat Rev Genet 7:395–401

Richards CL, Bossdorf O, Verhoeven KJ (2010) Understanding natural epigenetic variation. New Phytol 187:562–564

Schultz RJ (1980) Role of polyploidy in the evolution of fishes. In: Lewis WH (ed) Polyploidy: biological relevance. Plenum Press, New York, pp 313–340

Segraves KA, Thompson JN (1999) Plant polyploidy and pollination: floral traits and insect visits to diploid and tetraploid *Heuchera grossulariifolia*. Evolution 53:1114–1127

Semel Y et al (2006) Overdominant quantitative trait loci for yield and fitness in tomato. Proc Natl Acad Sci U S A 103:12981–12986

Sexton OJ (1980) Polyploidy in animal evolution: summary. See Ref. 69, pp 379–381

Shahbazian MD, Grunstein M (2007) Functions of site-specific histone acetylation and deacetylation. Annu Rev Biochem 76:75–100

Shen H, He H, Li J, Chen W, Wang X, Guo L, Peng Z, He G, Zhong S, Qi Y, Terzaghi W, Deng XW (2012) Genome-wide analysis of DNA methylation and gene expression changes in two Arabidopsis ecotypes and their reciprocal hybrids. Plant Cell 24:875–892

Shull GF (1908) The composition of a field of maize. Rep Am Breed Assoc 5:51–59

Shull GF (1952) Beginnings of the heterosis concept. In: Gowen JW (ed) Heterosis. Iowa State College Press, Ames, pp 14–48

Stebbins GL (1950) Variation and evolution in plants. Columbia University Press, New York

Stebbins GL (1980) Polyploidy: future prospects. In: Lewis WH (ed) Polyploid—biological relevance. Plenum, New York, pp 495–520

Stuber CW (1999) Biochemistry, molecular biology, and physiology of heterosis. In: Coors JG, Pandey S (eds) The genetics and exploitation of heterosis in crops. American Society of Agronomy, Inc/Crop Science Society of America, Inc, Madison, pp 173–184

Stupar RM, Springer NM (2006) Cis-transcriptional variation in maize inbred lines B73 and Mo17 leads to additive expression patterns in the F1 hybrid. Genetics 173:2199–2210

Swanson-Wagner RA et al (2006) All possible modes of gene action are observed in a global comparison of gene expression in a maize F1 hybrid and its inbred parents. Proc Natl Acad Sci U S A 103:6805–6810

Tate JA, Soltis DE, Soltis PS (2005) Polyploidy in plants. In: Gregory TR (ed) The evolution of the genome. Elsevier, San Diego, pp 371–426

Yu SB, Li JX, CG X, Tan YF, Gao YJ, Li XH, Zhang Q, Saghai Maroof MA (1997) Importance of epistasis as the genetic basis of heterosis in an elite rice hybrid. Proc Natl Acad Sci U S A 94:9226–9231

Zhang X, Yazaki J, Sundaresan A, Cokus S, Chan SW, Chen H, Henderson IR, Shinn P, Pellegrini M, Jacobsen SE et al (2006) Genome-wide high-resolution mapping and functional analysis of DNA methylation in Arabidopsis. Cell 126:1189–1201

Zhang HY, He H, Chen LB, Li L, Liang MZ, Wang XF, Liu XG, He GM, Chen RS, Ma LG, Deng XW (2008) A genome-wide transcription analysis reveals a close correlation of promoter INDEL polymorphism and heterotic gene expression in rice hybrids. Mol Plant 1:720–731

Zhao XX, Chai Y, Liu B (2007) Epigenetic inheritance and variation of DNA methylation level and pattern in maize intra-specific hybrids. Plant Sci 172:930–938

Zhu JK (2009) Active DNA demethylation mediated by DNA glycosylases. Annu Rev Genet 43:143–166

Zilberman D, Gehring M, Tran RK, Ballinger T, Henikoff S (2007) Genome-wide analysis of *Arabidopsis thaliana* DNA methylation uncovers an interdependence between methylation and transcription. Nat Genet 39:61–69

Zirkle C (1952) Early ideas on inbreeding and crossbreeding. In: Gowen JW (ed) Heterosis. Iowa State College Press, Ames, pp 1–13

Polyploidy and Genomic Changes

6

6.1 Introduction

Doubling of two different genomes or a single genome within a cell requires a series of genetic and genomic adjustments. Recent studies have shown that polyploid genomes may undergo rapid changes in structure and function through genetic and epigenetic variations (Osborn et al. 2003; Levy and Feldman 2004; Adams and Wendel 2005; Chen 2007; Zhang et al. 2009; Buggs et al. 2010a, b, 2011; Xiong et al. 2011; Matsushita et al. 2012; Wang et al. 2013; Dar et al. 2013; Dar et al. 2017). These can be dealt with under the following headings.

6.2 DNA Loss

The major fraction of a eukaryotic genome contains non-coding DNA sequences. Recent researches have shown that these DNA sequences may be lost in some species following polyploid formation (Ozkan et al. 2001; Shaked et al. 2001; Kashkush et al. 2002; Han et al. 2005). Differential gene loss following polyploidy has been found in some duplicate genes (Clark et al. 2006; Grover et al. 2007; Buggs et al. 2010a, b; Khasdan et al. 2010). Song et al. (1995) observed changes in the genomic organization of synthetic allotetraploid, *Brassica* sp., and detected the loss of parental genome fragments. Although the mechanism of DNA loss is not well understood, unequal homologous recombination and illegitimate recombination are thought to be the main factors responsible for this loss (Leitch and Bennett 2004; Gaeta et al. 2007). Elimination of low-copy, non-coding DNA fragments has been reported in *Aegilops* and *Triticum* hybrids by using AFLP (Feldman et al. 1997; Ozkan et al. 2001; Shaked et al. 2001; Han et al. 2003). Change in low-copy coding DNA sequences has also been observed in newly synthesized amphiploids of *Triticum* and *Aegilops* (Liu et al. 1998). Using FISH, elimination of a parent-specific DNA repeat in wheat was investigated by Han et al. (2005). They used parent-specific, tandem DNA repeat as probe on three sets of newly synthesized

T.-Ul.-H. Dar, R.-Ul. Rehman, *Polyploidy: Recent Trends and Future Perspectives*, https://doi.org/10.1007/978-81-322-3772-3_6

allopolyploids of wheat. Drastic physical elimination of parental repetitive DNA was found in all the three new allopolyploids in the second and third generations. In a similar study using GISH and AFLP, Tu et al. (2009) found chromosomal elimination and genomic rearrangements in partial hybrids between *Brassica rapa* and *Isatis indigotica.* They also inferred that partial chromosomal elimination and genomic rearrangements are responsible for partial hybrids in this cross. However, no DNA loss was found in *Gossypium* (Liu et al. 2001; Adams and Wendel 2005) and in *Spartina anglica* allotetraploids (Baumel et al. 2002; Ainouche et al. 2004).

By using cDNA-AFLP in newly synthesized wheat allotetraploid, Kashkush et al. (2002) found that the gene loss included rRNA genes and those involved in metabolism, disease resistance and cell-cycle regulation. Comparative genomics of *Brassica oleracea* and *Arabidopsis thaliana* revealed that the protein-coding genes present in the parental genomes were lost after polyploidization (Town et al. 2006). Lukens et al. (2006) investigated DNA sequence changes in *Brassica napus* using RFLP in the first generation of polyploidy and concluded that rare genetic changes occur in S_0 polyploid generation. Khasdan et al. (2010) reported tissue-specific DNA sequence elimination in wheat following allopolyploidization.

Using AFLP and flow cytometry, Ozkan et al. (2006) also observed no evidence for DNA loss due to autopolyploidization in *Arabidopsis thaliana*. This suggests that it may not be a universal response to polyploid formation (Leitch and Bennett 2004). Rapid DNA loss was observed in the first and subsequent generations of autotetraploid *Phlox drummondii* (Raina et al. 1994), while *Elymus elongatus* showed a loss of 10% DNA compared with their diploid progenitors (Eilam et al. 2009). Synthetic autotetraploid *Paspalum notatum* showed 9.5% sequence elimination shortly after polyploidization (Martelotto et al. 2007). In autopolyploid *Phlox drummondii* by using AFLP, loss of DNA sequences was also observed after few generations of autopolyploidization (Dar et al. 2013).

6.3 Genomic Rearrangements

Genomic rearrangements occur throughout the lifespan of an established polyploid, which may promote its long-term diversification and existence. Chromosomal rearrangements have been observed in many allopolyploids, such as *Arabidopsis*, *Spartina*, *Tragopogon*, *Brassica* and wheat (Feldman et al. 1997; Zhao et al. 1998; Shaked et al. 2001; Pontes et al. 2004; Salmon et al. 2005; Tate et al. 2006; Gaeta et al. 2007; Bento et al. 2008; Hegarty and Hiscock 2008; Lim et al. 2008; Cheung et al. 2009), and may arise as a consequence of colinearity of gene order between closely related parental genomes, leading to potential mispairing between highly homoeologous chromosomes during meiosis (Gale et al. 1998; Hegarty and Hiscock 2008).

Molecular cytogenetic techniques, such as FISH and GISH, are used for analysing genomic rearrangements in polyploids. By employing FISH using 45 s rRNA and 5 s rRNA genes as probes, Pontes et al. (2004) inferred that the genome restructuring occurs in both natural and newly synthesized *Arabidopsis suecica*. Numerous chromosomal rearrangements have been detected by genomic in situ hybridization (GISH). Inter-genomic rearrangements have been observed in tetraploid tobacco (Kenton et al. 1993) and *Milium* (Bennett et al. 1992) by using this technique. In newly formed wheat allopolyploid, both genome-specific and chromosome-specific DNA sequences have been eliminated shortly after its formation (Feldman et al. 1997; Ozkan et al. 2001).

Genomic sequence changes are believed to be the outcome of chromosomal translocations and nonreciprocal transpositions (Ma and Gustafson 2008). In *Hepatica nobilis*, loss of rDNA loci and asymmetrical relocation has been observed (Weiss-Schneeweiss et al. 2007). Genomic rearrangements have been observed in some grass species by using comparative genomics (Bennetzen 2007). In *Brassica*, Song et al. (1995) found that genomic rearrangements may occur very soon after polyploid formation. Extensive sequence-level divergence has been observed in *Brassica napus* and its progenitor species by comparative mapping (Cheung et al. 2009). By using multiple molecular markers, AFLP, RFLP, ISSR, IRAP and REMAP, Ma and Gustafson (2005, 2006, 2008) and Bento et al. (2008) found extensive genomic rearrangement in rye-specific fragments related to retrotransposons. Genomic shock is thought to be the main cause for genomic rearrangements in allopolyploids (McClintock 1984). Transposon activity within the genome can increase the likelihood of chromosome breakage and sequence amplification of gene duplication (Jin and Bennetzen 1994; Hegarty and Hiscock 2008).

6.4 DNA Methylation Changes

Many recent studies have helped provide evidence for frequent epigenetic changes in neo-polyploids (Shaked et al. 2001; Madlung et al. 2002; Adams et al. 2003; Wang et al. 2004; Salmon et al. 2005; Gaeta et al. 2007; Zhang et al. 2008; Hegarty et al. 2011; Mayfield et al. 2011; Lavania et al. 2012; Dar et al. 2013). Epigenetic changes do not involve alteration in DNA sequences but affect gene expression through several interrelated modifications, including DNA methylation, histone modification and chromatin packaging (Wolffe and Matzke 1999; Jenuwein and Allis 2001) (Fig. 6.1). DNA methylation status changes in neo-polyploids are one of the most investigated epigenetic mechanisms. In *Arabidopsis* and *Brassica* allopolyploids, the silenced genes were associated with high levels of DNA methylation and low levels of histone acetylation (Chen and Pikaard 1997). Changes in DNA methylation pattern can generate new phenotypic variants that could be targeted by selection (Kalisz et al. 2004; Salmon et al. 2008). Most of the DNA methylation changes are due to hybridization rather than genome doubling (Salmon and Ainouche 2010).

DNA methylation, especially methylation of cytosine, has been reported as one of the most common covalent modifications of DNA, and specific gene expression, cell differentiation, chromatin inactivation, embryo growth, replication timing and carcinogenesis can be caused by both epigenetic and mutagenic effects (Rossi et al. 1997; Pikaard 2000; Kato et al. 2003; Rapp and Wendel 2005; Lu et al. 2007). In

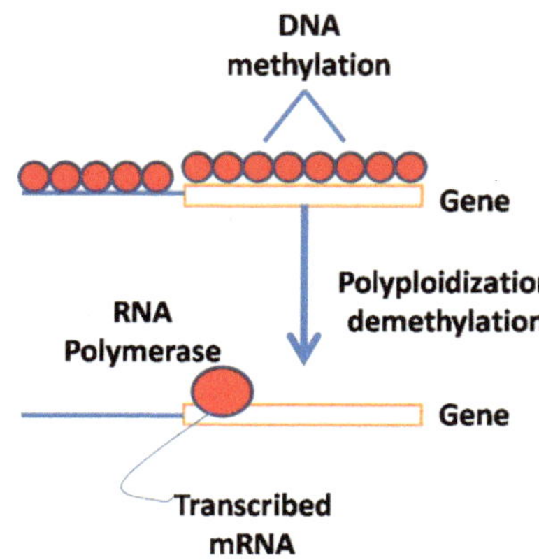

Fig. 6.1 Effect of polyploidy on epigenetic regulation. After genome doubling, a gene which has been silenced by DNA methylation in parents gets demethylated. As a result, RNA polymerase binds and starts transcription (Modified from Moghe and Shui 2014)

plants, cytosine methylation usually occurs at CpG and CpNG symmetric sequences, although cytosine methylation at non-symmetric sites is also common (Gruenbaum et al. 1981; Meyer et al. 1994; Pelissier et al. 1999). The methylation state can be maintained through cycles of DNA replication and is likely to play an integral role in regulating gene expression (Wassenegger 2000).

On polyploidization, methylation may result in control of the expression of redundant genes and genome stabilization (Lee and Chen 2001; Adams et al. 2003). For analysing DNA methylation, different techniques are used, such as High-performance Liquid Chromatography (HPLC), BisulphiteConversion, Methylation Specific PCR (MSP), Sequencing of Specific Gene, and Methylation Sensitive Amplification Polymorphism (MSAP). MSAP, combined with the appealing features of Amplified Fragment Length Polymorphism (AFLP), is based on the restriction enzyme and PCR amplification and is highly efficient for large-scale detection of cytosine methylation in the genome. MSAP has been extensively used to investigate methylation pattern changes in polyploids (Xiong et al. 1999; Xu et al. 2000; Cervera et al. 2002; Yi et al. 2005; Lu et al. 2007; Zilberman and Henikoff 2007; Jin et al. 2008; Chen et al. 2009a, b; Dar et al. 2013). On analysing the methylation status of the natural allopolyploid *Solanum ruiz-lealii* by MSAP technique and southern blotting, Marfil et al. (2009) found association between methylation pattern and abnormal flower phenotypes. By using MSAP, Verhoeven et al. (2010) observed methylation changes in new asexual triploid progeny in dandelion. These new triploid lineages in dandelion were formed by crossing diploid and triploid lines. Different methylation changes were observed in different progenies in each triploid lineage, generating novel DNA methylation variation between lineages that were mainly stably transmitted through generations (Salmon and Ainouche 2010).

Epigenetic alleles or epialleles differ from each other in the patterns of DNA methylation in the gene and affect ecologically important traits, such as flower shape, vegetative and seed pigmentation and pathogen resistance (Cubas et al. 1999; Chandler et al. 2000; Stokes and Richards 2002; Kalisz et al. 2004). Naturally occurring epialleles have also been found. Radially symmetric floral mutants of toad flax, *Linaria vulgaris*, exist in nature, instead of the wild-type bilaterally symmetric flowers (Cubas et al. 1999). Molecular data confirms that mutant (radial forms) results from epialleles of the CYCLOIDEA gene, which encodes a transcriptional activation, causing the development of floral asymmetry (Cubas et al. 1999). Naturally occurring epiallellic changes in maize result in altered seed and vegetative tissue pigmentation (Chandler et al. 2000). In *Arabidopsis*, abnormalities can be developed due to loss or gain of methylation-specific genes (FWA, SUP) (Jacobsen et al. 2000; Soppe et al. 2000).

6.5 Transposable Elements

Transposable elements represent about 50–80% of plant genomes (Feschotte and Pritham 2007) and play an important role in polyploidy evolution (Bennetzen 2000, 2005; Leitch and Leitch 2008; Parisod et al. 2010). Some recent studies confirm that polyploidy may lead to activation of transposons and retrotransposons (Liu and Wendel 2000; Kashkush et al. 2002, 2003; Madlung et al. 2002, 2005; Hegarty and Hiscock 2008; Dar et al. 2013). Allopolyploidy leads to increased transcription of particular retrotransposons (Waugh O'Neill et al. 1998; Labrador et al. 1999). Gerstel and Burns (1966) showed that interspecific hybrids between *Nicotiana otophora* and *N. tabacum* lead to the formation of megachromosomes up to 20–30-folds longer due to massive heterochromatin expansion. Later Comai (2000) suggested that this is due to activation of transposable elements (Leitch and Bennett 2004). Novel retrotransposon transcripts have been identified in polyploidy *Aegilops-Triticum* that were not present in diploid parents using cDNA-AFLP (Kashkush et al. 2002) in resynthesized allopolyploid *Nicotiana tabacum*. In a similar study, Parisod et al. (2010) found a few new transposable element insertions in *Spartina anglica*. It has been shown that activation of wis2-1A retrotransposons in wheat changes expression of adjacent genes (Kashkush et al. 2003). Madlung et al. (2005), using tilling microarray analysis, found that Ty-1 copia-like retrotransposons and two En/spm-like transposons belonging to sunfish (Suf) subfamily are reactivated in synthetic *Arabidopsis* allotetraploids. Epigenetic changes, such as DNA methylation, affect transposable element expression. In synthetic allopolyploid wheat and *Arabidopsis*, increased expression of transposable elements was found but was not always associated with increased transposition (reviewed in Parisod et al. 2010). Activation of retrotransposons in polyploids is attributed to reduction of methylation. For example, several cases of genome-wide methylation changes have been observed in newly synthesized wheat, in which retrotransposons were found to be demethylated (Shaked et al. 2001). Gypsy- and copia-like retrotransposons were found transcriptionally active, and their genomic copy number increased in wide synthetic hybrids from *Oryza sativa* and its wild relative *Zizania latifolia*. After a few generations, strong transcriptional suppression was observed (Liu and Wendel 2000). Molecular cytogenetic studies in cotton revealed that different transposable elements spread from one genome to another subgenome after allopolyploidization (Zhao et al. 1998).

After polyploidization, transposable elements get accumulated until full diploidization is reached. Transposable elements may have played a major role during structural diploidization in *Nicotiana* and maize. Reorganization of transposable element sequences immediately suggests that they play a central role during the early stage of allopolyploid evolution (reviewed in Parisod et al. 2010). Waugh et al. (1997) for the first time used transposon display approach, sequence-specific amplification polymorphism (SSAP), for revealing genetic distribution of Bare-1-like retrotransposable elements in the barley genome. Using this approach, Petit et al. (2010) detected amplification of Tnt1 retrotransposon. Polyploid-specific SSAP are often considered as indicator of transposition and are not present in the parental diploids.

6.6 Gene Expression Changes

Recent studies indicate that gene expressions in allopolyploids change with respect to their diploid progenitors and that the genes duplicated by polyploidy can have different expression patterns (Comai 2000; Kashkush et al. 2002; Adams et al. 2003; Wang et al. 2004; Madlung et al. 2005; Udall and Wendel 2006; Adams 2007; Hegarty et al. 2008; Gaeta et al. 2009; Jackson and Chen 2009; Coate and Doyle 2010; Buggs et al. 2010a, b; Wang et al. 2013). Gene expression changes occur immediately after allopolyploid formation or stochastically in the selfing progeny (Chen et al. 1998; Wang et al. 2004). It takes 1–2 generations for some genes, for example, rRNA genes, to establish a differential expression pattern and 2–5 generations for others to establish expression status (Chen and Ni 2006). Gene expression in polyploids may be either additive or nonadditive. In additive gene expression, the expression value of a gene in an allotetraploid is equal to 2, that is, (1 + 1 = 2). If the gene expression is nonadditive, the value would be either larger or less than 2. Both additive and nonadditive gene expressions have been observed in a variety of allopolyploids, such as *Arabidopsis*, cotton, wheat and *Senecio*. In *Arabidopsis*, cotton and *Senecio*, additive gene expression proportion varies from 65 to 95%, 30–70% and 40–73%, respectively. Nonadditive gene expression in wheat, *Senecio* and *Arabidopsis* includes 30–70%, 30–60% and 5–38%, respectively (reviewed in Jackson and Chen 2010). In *Arabidopsis* and *Brassica*, the gene involved in regulating flowering time, flowering locus C (FLC), displays additive effects on flowering time (Michaels and Amasino 1999), and also the expression in *Arabidopsis* allotetraploids, *A. thaliana* and *A. arenosa*, was found to be additive, giving rise to late flowering phenotype (Wang et al. 2006). Most of the changes observed in gene expression in early generations were found to be maintained in the late generations in allotetraploids. This suggested that rapid and stochastic changes are responsible for adaptive evolution in the resynthesized allotetraploids (Wang et al. 2004; Chen 2007). Gene expression increases as ploidy increases. In maize, expression of most genes increases in a euploid series (Guo et al. 1996). Allele-dosage effects in diploids have been observed in many important genes which are involved in developmental processes in plants. These include tb1 gene in maize (Doebley et al. 1997),

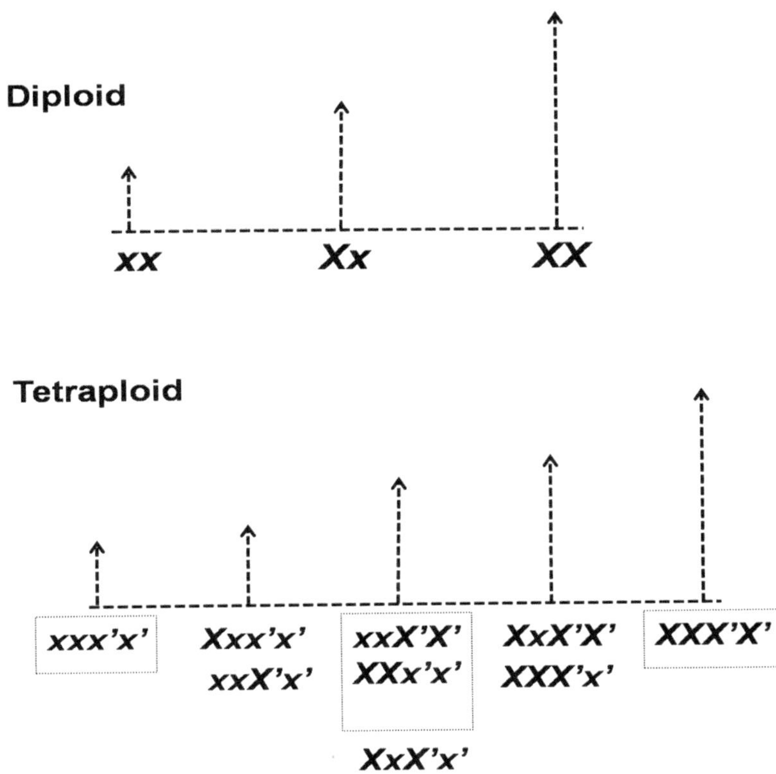

Fig. 6.2 Effect of dosage-regulated genes on polyploidy. In diploids, three genotypes with two alleles (X and x) expand to nine genotypes at the tetraploid level. Gene expression and phenotypic effects expand from diploid to tetraploid as shown by height of arrows (Modified from Osborn et al. 2003)

FLC in *Arabidopsis* (Michaels and Amasino 1999) and fruit size in tomato (FW2.2) (Frary et al. 2000) and are also involved in quantitative traits in general (Birchler et al. 2001; Guo and Birchler 1994). As compared to homozygous gene types which have high-expressing or low-expressing alleles, heterozygous genotypes have intermediate gene expression levels and phenotypic effects. Polyploidy increases the potential variation in expression levels for the genes having allele-dosage effects (Fig. 6.2).

Many genes involved in photochrome signalling, chlorophyll biosynthesis and starch metabolism are upregulated in *Arabidopsis* allotetraploids, and some are non-additively expressed in *Brassica*, *Senecio* and wheat allopolyploids. Some phytohormone and stress response genes have been seen to be nonadditively expressed in *Arabidopsis* allopolyploids. Stress-related genes are preferentially retained during polyploidization and show higher levels of expression divergence in *Arabidopsis thaliana* and in other resynthesized and natural allotetraploids (Ha et al. 2009). Small RNAs, including microRNAs (miRNAs), small interfering RNAs (siRNAs) and trans-acting siRNAs (tasiRNAs), have been found to mediate

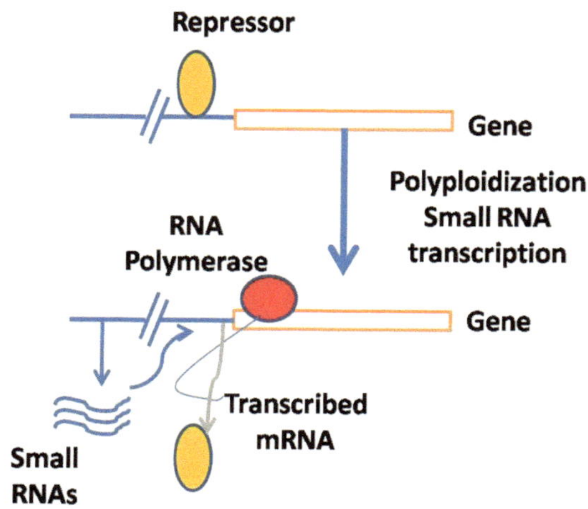

Fig. 6.3 Effect of polyploidy on regulation of small RNAs. In the parents, a gene is not expressed due to binding of repressor to its promotor. After polyploidization, the repressor is removed due to regulation by small RNAs and gene is expressed (Modified from Moghe and Shui 2014)

post-transcriptional regulation, RNA-directed DNA methylation and chromatin remodelling (reviewed in: Jackson and Chen 2010). Recently, Ha et al. (2009) found roles for small RNAs in maintaining genome expression in *Arabidopsis* allopolyploids. It was found that repeat associated siRNAs were highly diverged between *Arabidopsis thaliana* and *Arabidopsis arenosa*, and tasiRNAs mediate expression diversity between closely related species and in the allotetraploids (Fig. 6.3). Homoeologous genes express at different levels in allopolyploids and respond differently to allopolyploidy in various organs of the plant. Bottley et al. (2006) documented organ-specific silencing in wheat allohexaploids. Organ-specific variation in homoeologous gene expression can be an immediate response to allopolyploidy, as observed in neo-polyploids of *Gossypium* and *Arabidopsis* (Adams 2007).

Changes in chromosome morphology might also affect gene expression (McClintock 1984; Navasin 1934). Disappearance of secondary constrictions by silencing of rDNA loci has also been reported (Pikaard 1999). Nuclear dominance is thought to be associated with silencing of rDNA genes. cDNA-AFLP technique has been used to confirm differentially expressed genes in allopolyploid species, such as *Arabidopsis*, *Triticum*, *Brassica* and *Tragopogon* (Comai et al. 2000; Lee and Chen 2001; Kashkush et al. 2002; He et al. 2003; Soltis et al. 2004; Tate et al. 2006; Gaeta et al. 2007).

Using quantitative reverse transcriptase-polymerase chain reaction and high-throughput RNA sequencing, Coate and Doyle (2010) calculated relative size of leaf transcriptome. It was found that leaf transcriptome was 1.4-fold larger than either progenitor transcriptomes. This approach can be used to estimate relative expression per cell. With conjunction to transcript profiling data, it can also be used to compare the size of whole transcriptomes, which in turn makes possible to compare expression per cell for each gene in the transcript profiling data set (Coate and Doyle 2010).

Microarray technology offers valuable opportunities for genome-wide analysis of large number of genes in parallel and has been used to study polyploidization in resynthesized plant polyploids (Gaeta et al. 2009). Wang et al. (2006) analysed

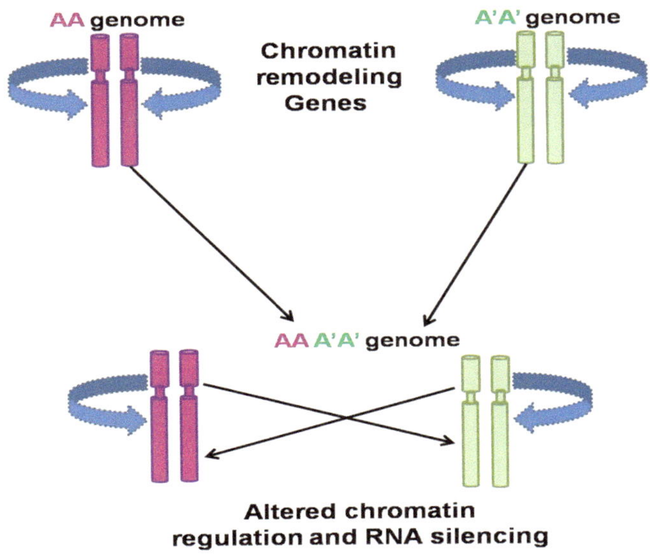

Fig. 6.4 This illustration shows the effect of polyploidy on chromatin structure and gene expression. AA and A′A′ represent two different genomes of diploids. Chromatin remodelling factors affect the chromatin only within AA and A′A′ genomes of diploids separately. In tetraploids (AAA′A′), chromatin remodelling factors can affect the chromatin of both genomes. This in turn can cause chromatin remodelling and RNA silencing on new polyploids (Modified from Osborn et al. 2003)

transcriptome divergence in resynthesized *Arabidopsis* allopolyploids and reported nonadditive expression of 3.1% genes out of about 26,000 genes in two independent lineages. It was found that up to 38% of genes showed change in gene expression within lines. Using cDNA microarray approach, Hegarty et al. (2008) found nonadditive and transgressive gene expression changes in *Senecio* hybrids. Recently, Gaeta et al. (2009) used 70 mer oligo-microarray to analyse gene expression in resynthesized *Brassica napus* allopolyploids and found nonadditive expression in a few genes. These authors also found that homoeologous rearrangements were not responsible for genome-wide nonadditive gene expression. In another study, Hegarty et al. (2006) did not find any change in expression in particular functional category of genes, although genes involved in floral or pollen development were slightly overrepresented. Similarly, by using microarray, *Helianthus deserticola* hybrid showed differential regulation of genes in various functional categories (Lai et al. 2006).

In allopolyploids, the role of chromatin modifications in silencing or activation of protein-coding genes has been demonstrated in several recent studies (Lee and Chen 2001; Madlung et al. 2002; Kashkush et al. 2003; Wang et al. 2004, 2006). Silenced genes can be reactivated by aza-dc, a chemical inhibitor of DNA methylation (Lee and Chen 2001) or by down-regulation of the genes encoding DNA methyltransferase using RNA interference (RNAi) (Wang et al. 2004). Histone modifications, such as acetylation, phosphorylation, ubiquitination and methylation, have been associated recently with heterochromatic silencing (Jenuwein and Allis 2001) (Fig. 6.4).

Much of the information discussed above has been obtained from allopolyploids, only a few reports being on the changes in gene expression due to autopolyploidization (Guo et al. 1996; Galitski et al. 1999; Albertin et al. 2005; Auger et al. 2005; Lu et al. 2006; Storchova et al. 2006; Church and Spaulding 2009). Using cDNA-AFLP, Church and Spaulding (2009) suggested that gene expression is not affected by genome doubling in the long term. Albertin et al. (2005) found no significant changes in gene expression in synthetic autopolyploids of *Brassica oleracea*, by investing its proteome. Stupar et al. (2007) developed synthetic autopolyploid series (1x–4x) in potato (*Solanum phureja*) to investigate transcriptomic changes by using cDNA microarray. Nearly about 10% of genes out of about 9000 showed quantitative expression changes among ploidy levels; most changes were associated with the monoploid up to the twofold level.

Next-generation sequencing (NGS) technologies are high throughput employed for RNA sequencing which are directly sequenced to determine their base composition. RNA sequencing is more advantageous than hybridization-based RNA quantification techniques because its data gives an absolute number of reads for each gene. Thus these are more suitable for gene expression analysis having less error to allow determination of absolute expression levels (Wang et al. 2009). In *Tragopogon miscellus*, allopolyploids NGS has permitted the genome-wide expression of small RNAs. While studying the differential expression of duplicate homologous genes derived from parental genes by profiling about 3000 SNPs using Illumina RNA sequencing approach, it was found that there were expression biases between diploids and natural allopolyploids and uniform expression in F1 and F1 synthetic allopolyploids. Explaining the above 'transcriptomic shock', it was hypothesized that loosening of gene expression regulation results in gradual evolution in early generations of polyploids (Buggs et al. 2010a, b). Employing qRT PCR and RNA sequencing in *Glycine dolichocarpa* and its diploid progenitors, it is found that most genes had partial dosage compensation (Coate and Doyle 2010). Further, the RNA sequencing in *Glycine max* identified likely genes showing the differences in photosynthetic rates between allotetraploids and its progenitors (Ilut et al. 2012). Furthermore, the transcriptomic profiling in *A. thaliana* carried out by cDNA pyrosequencing suggesting that small RNAs have a role in buffering the genomic shock in interspecific hybrids and allopolyploids, more seemingly in maintaining the stability of genome and chromatin and modulating the nonadditive gene expression. It has also been found in *A. thaliana*, *A. arenosa* and their allopolyploids that the repeat- and transposon-associated siRNAs are divergent and their nonadditive gene expressions were not correlated. However, the miRNA and tasiRNA sequences are reported to be conserved with variable expression patterns in allotetraploids and their progenitors (Ha et al. 2009).

6.7 Allelic Sub-functionalization and Neo-functionalization

Sub-functionalization (functional divergence), which refers to the partitioning of ancestral functions or expression patterns between duplicated genes so that both are retained, is one of the major outcomes for duplicated genes (Force et al. 1999; Blanc and Wolfe 2004; Adams 2007). After genome doubling, both sub-functionalization and neo-functionalization play an important role in the functioning and diversification of duplicate genes. Organ- or tissue-specific expression patterns of duplicate genes were first observed in allopolyploid *Brassica* and *Gossypium* (Chen et al. 2007). Sub-functionalization in different plant organs has now been reported in a number of plants (Adams et al. 2003; Cusack and Wolfe 2007; Chaudhary et al. 2009). Adams et al. (2003) found organ-specific regulation of alcohol dehydrogenase gene Adh A in various floral organs of cotton allopolyploid, with one homoeolog showing silencing in some organs and the other being silenced in other organs. Blanc and Wolfe (2004), using functional genomics data for a large number of duplicated gene pairs formed by ancient polyploidy events in *Arabidopsis thaliana*, inferred that functional diversification of the surviving duplicated genes is a major feature of the long-term evolution of polyploids (Fig. 6.5).

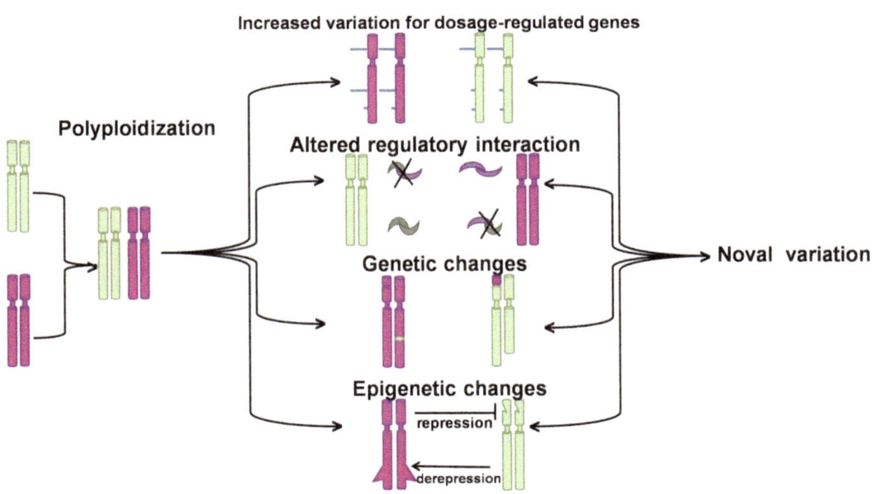

Fig. 6.5 Factors responsible for novel variation in polyploids. Combination of two different genomes (indicated with two different colours) from two diploid parents can cause:
Increased variation for dosage-regulated genes affects the magnitude of allelic expression. This effect is shown by size of blocks for three different loci.
Altered regulatory interaction. Regulatory factors (trans-acting) shown as dimeric protein. Heterodimers not functioning properly.
Genetic changes such as insertion, deletion, conversion and translocation, all these affect gene expression.
Epigenetic changes such as repression and depression which are caused by merging of two different genes can lead to change in gene expression which in turn affects transposable elements (Modified from Osborn et al. 2003)

Sub-functionalization of ribosomal proteins in *Brassica napus* has been recently observed by Whittle and Krochko (2009). Transcript profiling reveals that large fractions of ribosomal protein genes were differentially expressed and that the number of paralogs expressed for each ribosomal protein varies extensively in different tissues of *Brassica napus*. Novel ribosomal protein transcripts were also observed in each tissue. By using a novel approach of genome-specific mass spectrometry, Chaudhary et al. (2009) investigated relative levels of allelic and homoeologous gene expression in diploid and allopolyploid cotton. Transcriptional sub-functionalization was observed in most genes, for example, gene CO131164 was found silenced in the ovary wall of A-genome homoeolog, but its expression was observed to be 93% in anthers of A-genome homoeolog of allopolyploid cotton.

Maize genes PhyB1 and PhyB2 (Sheehan et al. 2007) and OSMADS3 and OSMADS58 genes in rice (Yamaguchi et al. 2006) are the examples of sub-functionalization in plants, where partitioning of function occurs between duplicate genes. The Organ-specific silencing of homologs has also been observed in wheat allohexaploid (Bottley et al. 2006). Sub-functionalization occurs immediately after polyploid formation, as has been observed in neo-polyploids of *Gossypium* and *Arabidopsis*. *Gossypium* neo-polyploids show considerable variation in the expression level of homoeologous gene pairs in different floral organs (Adams et al. 2004). Cusack and Wolfe (2007), on the basis of functional genomics, confirmed that the chloroplast (RPL32) and nuclear (SODCP) genes integrated in ancestors of mangrove and poplar trees, forming bifunctional gene by alternative splicing. This bifunctional gene was retained in mangroves but underwent sub-functionalization to reform separate RPL32 and SODCP genes in poplar.

Neo-functionalization, the gain of new function and expression patterns, has been observed in many plants, including lectins in legumes (van Damme et al. 2007) and MADS-box genes in *Physalis* (He and Saedler 2005), *Gossypium* (Chaudhary et al. 2009) and *Arabidopsis* (Duarte et al. 2006). Neo-functionalization in eight different tissues in allopolyploid cotton has been observed by Chaudhary et al. 2009. Genes CO108066 and CO076921 failed to show expression in either the A or D orthologs of newly formed synthetic allopolyploid cotton, respectively, in leaf lamina. But both homoeologs showed expression in the same tissue in natural allopolyploid.

References

Adams KL (2007) Evolution of duplicate gene expression in polyploid and hybrid plants. J Hered 98:136–141

Adams KL, Wendel JF (2005) Allele-specific, bidirectional silencing of an alcohol dehydrogenase gene in different organs of interspecific diploid cotton hybrids. Genetics 171:2139–2142

Adams KL, Cronn R, Percifield R, Wendel JF (2003) Genes duplicated by polyploidy show unequal contributions to the transcriptome and organ-specific reciprocal silencing. Proc Natl Acad Sci U S A 100(8):4649–4654

Adams KL, Percifield R, Wendel JF (2004) Organ-specific silencing of duplicated genes in a newly synthesized cotton allotetraploid. Genetics 168:2217–2226

Ainouche ML, Baumel A, Salmon A (2004) *Spartina anglica* C.E. Hubbard: a natural model system for analysing early evolutionary changes that affect allopolyploid genomes. Biol J Linn Soc 82:475–484

Albertin W, Brabant P, Catrice O, Eber F, Jenczewski E, Chevre AM, Thiellement H (2005) Autopolyploidy in cabbage (Brassica oleracea L.) does not alter significantly the proteomes of green tissues. Proteomics 5:2131–2139

Auger DL, Gray AD, Ream TS, Kato A, Coe EH, Birchler JA (2005) Nonadditive gene expression in diploid and triploid hybrids of maize. Genetics 169:389–397

Baumel A, Ainouche M, Kalendar R, Schulman AH (2002) Retrotransposons and genomic stability in populations of the young allopolyploid species *Spartina anglica* C.E. Hubbard (Poaceae). Mol Biol Evol 19:1218–1227

Bennett ST, Kenton AY, Bennett MD (1992) Genomic in situ hybridization reveals the allopolyploid nature of *Milium montianum* (Gramineae). Chromosoma 101:420–424

Bennetzen JL (2000) Transposable element contributions to plant gene and genome evolution. Plant Mol Biol 42:251–269

Bennetzen JL (2005) Transposable elements, gene creation and genome rearrangement in flowering plants. Curr Opin Genet Dev 15:621–627

Bennetzen JL (2007) Patterns in grass genome evolution. Curr Opin Plant Biol 10:176–181

Bento M, Pereira HS, Rocheta M, Gustafson P, Viegas W, Silva M (2008) Polyploidization as a retraction force in plant genome evolution: sequence rearrangements in triticale. PLoS One 3:e1402

Birchler JA, Bhadra U, Pal-Bhadra M, Auger DL (2001) Dosage dependent gene regulation in higher eukaryotes: implications for dosage compensation, aneuploid syndromes and quantitative traits. Dev Biol 234:275–288

Blanc G, Wolfe KH (2004) Functional divergence of duplicated genes formed by polyploidy during Arabidopsis evolution. Plant Cell 16:1679–1691

Bottley A, Xia G, Koebner R (2006) Homoeologous gene silencing in hexaploid wheat. Plant J 47:897. Proc Natl Acad Sci U S A 100:4649–4654

Buggs RJ, Chamala S, Wu W, Gao L, May GD, Schnable PS, Soltis DE, Soltis PS, Barbazuk WB (2010a) Characterization of duplicate gene evolution in the recent natural allopolyploid *Tragopogon miscellus* by next-generation sequencing and Sequenom iPLEX MassARRAY genotyping. Mol Ecol 19:S132–S146

Buggs RJ, Elliott NM, Zhang L, Koh J, Viccini LF, Soltis DE et al (2010b) Tissue-specific silencing of homoeologs in natural populations of the recent allopolyploid *Tragopogon mirus*. New Phytol 186:175–183

Buggs RJA, Zhang L, Miles N, Tate JA, Gao L, Wei W et al (2011) Transcriptomic shock generates evolutionary novelty in a newly formed, natural allopolyploid plant. Curr Biol 21:551–556

Cervera MT, Ruiz-Garcia L, Martinez-Zapater JM (2002) Analysis of DNA methylation in *Arabidopsis thaliana* based on methylation-sensitive AFLP markers. Mol Genet Genomics 268:543–552

Chandler VL, Eggleston WB, Dorweiler JE (2000) Paramutation in maize. Plant Mol Biol 43:121–145

Chaudhary B, Flagel L, Stupar RM, Udall JA, Verma N, Springer NM, Wendel JF (2009) Reciprocal silencing, transcriptional bias and functional divergence in homeologs in polyploid cotton (Gossypium). Genetics 182:503–517

Chen ZJ (2007) Genetic and epigenetic mechanisms for gene expression and phenotypic variation in plant polyploids. Annu Rev Plant Biol 58:377–406

Chen ZJ, Ni Z (2006) Mechanisms of genomic rearrangements and gene expression changes in plant polyploids. BioEssays 28:240–252

Chen ZJ, Pikaard CS (1997) Transcriptional analysis of nucleolar dominance in poly-ploid plants: biased expression/silencing of progenitor rRNA genes is developmentally regulated in Brassica. Proc Natl Acad Sci U S A 94:3442–3447

Chen ZJ, Comai L, Pikaard CS (1998) Gene dosage and stochastic effects determine the severity and direction of uniparental ribosomal RNA gene silencing (nucleolar dominance) in Arabidopsis allopolyploids. Proc Natl Acad Sci U S A 95:14891–14896

Chen ZJ, Ha M, Soltis D (2007) Polyploidy: genome obesity and its consequences. New Phytol 174(4):717–720

Chen X, Ma Y, Chen F, Song W, Zhang L (2009a) Analysis of DNA methylation patterns of PLBs derived from *Cymbidium hybridium* based on MSAP. Plant Cell Tissue Organ Cult 98:67–77

Chen X, Wang XJ, Chen ZJ (2009b) Small RNAs serve as a genetic buffer against genomic shock in Arabidopsis interspecific hybrids and allopolyploids. Proc Natl Acad Sci U S A 106:17835–17840

Cheung F, Trick M, Drou N, Lim YP, Park J-Y, Kwon S-J, Kim J-A, Scott R, Pires JC, Paterson AH et al (2009) Comparative analysis between homoeologous genome segments of Brassica napus and its progenitor species reveals extensive sequence-level divergence. Plant Cell 21:1912–1928

Church SA, Spaulding EJ (2009) Gene expression in a wild autopolyploid sunflower series. J Hered 100(4):491

Clark RM, Wagler TN, Quijada P, Doebley J (2006) A distant upstream enhancer at the maize domestication gene tb1 has pleiotropic effects on plant and inflorescent architecture. Nat Genet 38:594–597

Comai L (2000) Genetic and epigenetic interactions in allopolyploid plants. Plant Mol Biol 43:387–399

Comai L, Tyagi AP, Winter K, Holmes-Davis R, Reynolds SH, Stevens Y, Byers B (2000) Phenotypic instability and rapid gene silencing in newly formed Arabidopsis allotetraploids. Plant Cell 12:1551–1567

Coate JE, Doyle JJ (2010) Quantifying whole transcriptome size, a prerequisite for understanding transcriptome evolution across species: an example from a plant allopolyploid. Genome Biol Evol 2:534–546

Cubas P, Vincent C, Coen E (1999) An epigenetic mutation responsible for natural variation in floral symmetry. Nature 401:157–161

Cusack BP, Wolfe KH (2007) When gene marriages don't work out: divorce by subfunctionalization. Trends Genet 23:270–272

Dar TH, Raina SN, Goel S (2013) Molecular analysis of genomic changes in synthetic autotetraploids *Phlox drummondii* Hook. Biol J Linn Soc 110:591–605

Dar TH, Raina SN, Goel S (2017) Cytogenetic and molecular evidences revealing genomic changes after autopolyploidization: a case study of synthetic autotetraploid *Phlox drummondii* hook. Physiol Mol Biol Plants. https://doi.org/10.1007/s12298-017-0445-8

Doebley J, Stec A, Hubbard L (1997) The evolution of apical dominance in maize. Nature 386:485–488

Duarte JM, Cui L, Wall PK, Zhang Q, Zhang X et al (2006) Expression pattern shifts following duplication indicative of subfunctionalization and neofunctionalization in regulatory genes of Arabidopsis. Mol Biol Evol 23:469–478

Eilam T, Anikster Y, Millet E, Manisterski J, Feldman M (2009) Genome size in natural and synthetic autopolyploids and in natural segmental allopolyploid of several Triticeae species. Genome 52:275–285

Feldman M, Liu B, Segal G, Abbo S, Levy AA, Vega JM (1997) Rapid elimination of low-copy DNA sequences in polyploid wheat: a possible mechanism for differentiation of homoeologous chromosomes. Genetics 147:1381–1387

Feschotte C, Pritham EJ (2007) DNA transposons and the evolution of eukaryotic genomes. Annu Rev Genet 41:331–368

Force A, Lynch M, Pickett FB, Amores A, Yan Y-L et al (1999) Preservation of duplicate genes by complementary, degenerative mutations. Genetics 151:1531–1545

Frary A, Nesbitt TC, Grandillo S, Knaap E, Cong B, Liu J, Meller J, Elber R, Alpert KB, Tanksley SD (2000) fw2.2: a quantitative trait locus key to the evolution of tomato fruit size. Science 99:85–88

Gaeta RT, Pires JC, Iniguez-Luy F, Leon E, Osborn TC (2007) Genomic changes in resynthesized *Brassica napus* and their effect on gene expression and phenotype. Plant Cell 19:3403–3417

Gaeta RT, Yoo SY, Pires JC, Doerge RW, Chen ZJ, Osborn TC (2009) Analysis of gene expression in Resynthesized *Brassica napus* allopolyploids using Arabidopsis 70mer OligoMicroarrays. PLoS One 4:e4760

Gale P, Young C, Stanfield G, Oakes D (1998) Development of a risk assessment for BSE in the aquatic environment. J Appl Microbiol 84:467–477

Galitski T, Saldanha AJ, Styles CA, Lander ES, Fink GR (1999) Ploidy regulation of gene expression. Science 285:251–254

Gerstel DU, Burns JA (1966) In: Darlington CA, Lewis KR (eds) Chromosomes today, vol 1. Plenum Press, New York, pp 41–56

Grover CE, Kim H, Wing RA, Paterson AH, Wendel JF (2007) Microcolinearity and genome evolution in the *AdhA* region of diploid and polyploid cotton (*Gossypium*). Plant J 50:995–1006

Gruenbaum Y, Naveh-Many T, Cedar H, Razin A (1981) Sequence specificity of methylation in higher plant DNA. Nature 292:860–862

Guo M, Birchler JA (1994) Trans-acting dosage effects on the expression of model gene systems in maize aneuploids. Science 266:1999–2002

Guo M, Davis D, Bircher JA (1996) Dosage effects on gene expression in a maize ploidy series. Genetics 142:1349–1355

Ha M, Lu J, Tian L, Ramachandran V, Kasschau KD, Chapman EJ, Carrington JC, Chen X, Wang XJ, Chen ZJ (2009) Small RNAs serve as a genetic buffer against genomic shock in *Arabidopsis* interspecific hybrids and allopolyploids. Proc Natl Acad Sci U S A 106:17835–17840

Han FP, Fedak G, Ouellet T, Liu B (2003) Rapid genomic changes in interspecific and intergeneric hybrids and allopolyploids of *Triticeae*. Genome 46:716–723

Han F, Fedak G, Guo W, Liu B (2005) Rapid and repeatable elimination of a parental genome-specific DNA repeat (pGc1R-1a) in newly synthesized wheat allopolyploids. Genetics 170:1239–1245

He C, Saedler H (2005) Heterotopic expression of MPF2 is the key to the evolution of the Chinese lantern of *Physalis*, a morphological novelty in Solanaceae. Proc Natl Acad Sci U S A 102:5779–5784

He P, Friebe BR, Gill BS (2003) Allopolyploidy alters gene expressionin the highly stable hexaploid wheat. Plant Mol Biol 52:401–414

Hegarty MJ, Hiscock SJ (2008) Genomic clues to the evolutionary success of polyploid plants. Curr Biol 18:435–444

Hegarty MJ, Barker GL, Wilson ID, Abbott RJ, Edwards KJ et al (2006) Transcriptome shock after interspecific hybridization in senecio is ameliorated by genome duplication. Curr Biol 16:1652–1659

Hegarty MJ, Barker GL, Brennan AC, Edwards KJ, Abbott RJ, Hiscock SJ (2008) Changes to gene expression associated with hybrid speciation in plants: further insights from transcriptomic studies in Senecio. Philos Trans R Soc B 363:3055–3069

Hegarty MJ, Batstone TOM, Barker GL, Edwards KJ, Abbott RJ, Hiscock SJ (2011) Nonadditive changes to cytosine methylation as a consequence of hybridization and genome duplication in Senecio (Asteraceae). Mol Ecol 20:105–113

Ilut DC, Coate JE, Luciano AK, Owens TG, May GD, Farmer A, Doyle JJ (2012) A comparative transcriptomic study of an allotetraploid and its diploid progenitors illustrates the unique advantages and challenges of RNA-seq in plant species. Am J Bot 99:383–396

Jackson S, Chen ZJ (2009) Genomic and expression plasticity of polyploidy. Curr Opin Plant Biol 13:153–159

Jackson S, Chen ZJ (2010) Genomic and expression plasticity of polyploidy. Curr Opin Plant Biol 13:153–159

Jacobsen SE, Sakai H, Finnegan EJ, Cao X, Meyerowitz EM (2000) Ectopic hypermethylation of flower-specific genes in *Arabidopsis*. Curr Biol 10:179–186

Jenuwein T, Allis CD (2001) Translating the histone code. Science 293(5532):1074–1080

Jin YK, Bennetzen JL (1994) Integration and non-random mutation of a plasma-membrane proton atpase gene fragment within the Bs1 retroelement of maize. Plant Cell 6:102–106

Jin H, Hu W, Wei Z, Wan L, Li G, Tan G, Zhu L, He G (2008) Alterations in cytosine methylation and species-specific transcription induced by interspecific hybridization between *Oryza sativa* and *O. officinalis*. Theor Appl Genet 117:1271–1279

Kalisz S, Vogler DW, Hanley KM (2004) Context-dependent autonomous self-fertilization yields reproductive assurance and mixed mating. Nature 430:884–887

Kashkush K, Feldman M, Levy AA (2002) Gene loss, silencing and activation in a newly synthesized wheat allotetraploid. Genetics 160:1651–1659

Kashkush K, Feldman M, Levy AA (2003) Transcriptional activation of retrotransposons alters the expression of adjacent genes in wheat. Nat Genet 33:102–106

Kato MA, Bender MJ, Jacobsen SE, Kakutani T (2003) Role of CG and non-CG methylation in immobilization of transposons in Arabidopsis. Curr Biol 13:421–426

Kenton A, Parokonny AS, Gleba YY, Bennett MD (1993) Characterization of the *Nicotiana tabacum* L. genome by molecular cytogenetics. Mol Gen Genet 240:159–169

Khasdan V, Yaakov B, Kraitshtein Z, Kashkush K (2010) Developmental timing of DNA elimination following allopolyploidization in wheat. Genetics 185:387–390

Labrador M, Farre M, Utzet F, Fontdevila A (1999) Interspecific hybridization increases transposition rates of Osvaldo. Mol Biol Evol 16:931–937

Lai LC, Kosorukoff AL, Burke PV, Kwast KE (2006) Metabolic-state-dependent remodeling of the transcriptome in response to anoxia and subsequent reoxygenation in saccharomyces cerevisiae. Eukaryatic Cell 5(9):1468–1489

Lavania UC, Srivastava S, Lavania S, Basu S, Misra NK, Mukai Y (2012) Autopolyploidy differentially influences body size in plants, but facilitates enhanced accumulation of secondary metabolites, causing increased cytosine methylation. Plant J 71:539–549

Lee HS, Chen ZJ (2001) Protein-coding genes are epigenetically regulated in Arabidopsis polyploids. Proc Natl Acad Sci U S A 98:6753–6758

Leitch IJ, Bennett MD (2004) Biological relevance of polyploidy: ecology to genomics. Biol J Linn Soc 84:651–663

Leitch AR, Leitch IJ (2008) Genomic plasticity and the diversity of polyploid plants. Science 320:481–483

Levy AA, Feldman M (2004) Genetic and epigenetic reprogramming of the wheat genome upon allopolyploidization. Biol J Linn Soc 82:607–613

Lim KY, Soltis DE, Soltis PS, Tate J, Matyasek R, Srubarova H, Kovarik A, Pires JC, Xiong Z, Leitch AR (2008) Rapid chromosome evolution in recently formed polyploids in Tragopogon (Asteraceae). PLoS One 3:e3353

Liu B, Wendel JF (2000) Retrotransposon activation followed by rapid repression in introgressed rice plants. Genome 43:874–880

Liu B, Vega JM, Segal G, Abbo S, Rodova M, Feldman M (1998) Rapid genomic changes in newly synthesized amphiploids of Triticum and Aegilops I Changes in low-copy noncoding DNA sequences. Genome 41:272–277

Liu B, Brubaker CL, Mergeai G, Cronn RC, Wendel JF (2001) Polyploid formation in cotton is not accompanied by rapid genomic changes. Genome 44:321–330

Lu BB, Pan XZ, Zhang L, BBM H, Sun L, Bin L, Yi B, Zheng SQM, Yu XJ, Ding RX et al (2006) A genome-wide comparison of gene responsive to autopolyploidy in *Isatis indigotica* using *Arabidopsis thaliana* affymetrix genechips. Plant Mol Biol Report 24:197–204

Lu G, Wu X, Chen B, Gao G, Xu K (2007) Evaluation of genetic and epigenetic modification in rapeseed (*Brassica napus*) induced by salt stress. J Integr Plant Biol 49:1599–1607

Lukens LN, Pires JC, Leon E, Vogelzang R, Oslach L, Osborn T (2006) Patterns of sequence loss and cytosine methylation within a population of newly resynthesized *Brassica napus* allopolyploids. Plant Physiol 140:336–348

Ma XF, Gustafson JP (2005) Genome evolution of allopolyploids: a process of cytological and genetic diploidization. Cytogenet Genome Res 109:236–249

Ma XF, Gustafson JP (2006) Timing and rate of genome variation in triticale following allopolyploidization. Genome 49:950–958

Ma XF, Gustafson JP (2008) Allopolyploidization-accommodated genomic sequence changes in triticale. Ann Bot 101:825–832

Madlung A, Masuelli RW, Watson B, Reynolds SH, Davison J, Comai L (2002) Remodeling of DNA methylation and phenotypic and transcriptional changes in synthetic Arabidopsis allotetraploids. Plant Physiol 129:733–746

Madlung A, Tyagi AP, Watson B, Jiang HM, Kagochi T, Doerge RW, Martienssen R, Comai L (2005) Genomic changes in synthetic Arabidopsis polyploids. Plant J 41:221–230

Marfil CF, Camadro EL, Masuelli RW (2009) Phenotypic instability and epigenetic variability in a diploid potato of hybrid origin, *Solanum ruiz-lealii*. BMC Plant Biol 9:21

Martelotto LG, Ortiz JPA, Stein J, Espinoza F, Quarin CL, Pessino SC (2007) Genome rearrangements derived from autopolyploidization in Paspalum sp. Plant Sci 172:970–977

Matsushita SC, Tyagi AP, Thornton GM, Pires JC, Madlung A (2012) Allopolyploidization lays the foundation for evolution of distinct populations: evidence from analysis of synthetic Arabidopsis allohexaploids. Genetics 191:535–547

Mayfield D, Chen ZJ, Pires JC (2011) Epigenetic regulation of flowering time in polyploids. Curr Opin Plant Biol 14:174–178

McClintock B (1984) The significance of responses of the genome to challenge. Science 226:792

Meyer P, Niedenhof I, TenLohuis M (1994) Evidence for cytosine methylation of non-symmetrical sequences in transgenic Petunia hybrida. EMBO J 13:2084–2088

Michaels SD, Amasino RM (1999) Flowering locus C encodes a novel MADS domain protein that acts as a repressor of flowering. Plant Cell 11:949–956

Moghe GD, Shiu SH (2014) The causes and molecular consequences of polyploidy in flowering plants. Ann N Y Acad Sci 1320:16–34

Navasin M (1934) Chromosomal alteration caused by hybridization and their bearing upon certain general genetic problem. Cytologia 5:169–203

Osborn TC, Pires JC, Birchler JA, Auger DL, Chen ZJ, Lee HS, Comia L, Madlungh A, Doerge RW, Colot V, Matrienssen RA (2003) Understanding mechanisms of novel gene expression in polyploids. Trends Genet 19:141–147

Ozkan H, Levy AA, Feldman M (2001) Allopolyploidy induced rapid genome evolution in the wheat (Aegilops – Triticum) group. Plant Cell 13:1735–1747

Ozkan H, Tuna M, Galbraith DW (2006) No DNA loss in autotetraploids of Arabidopsis thaliana. Plant Breed 125:288–291

Parisod C, Alix K, Just J, Petit M, Sarilar V, Mhiri C, Ainouch M, Chalhoub B, Grandbastien MA (2010) Impact of transposable elements on the organization and function of allopolyploid genomes. New Phytol 186:37–45

Pelissier T, Thalmier S, Kempe D, Sanger HL, Wassenegger M (1999) Heavy de novo methylation and non-symmetrical sites is a hallmark of RNA-directed DNA methylation. Nucleic Acids Res 27:1625–1634

Petit M, Guidat C, Daniel J, Denis E, Montoriol E, Bui QT, Lim KY, Kovarik A, Leitch AR, Grandbastien M-A, Mhiri C (2010) Mobilization of retrotransposons in synthetic allotetraploid tobacco. New Phytol 186:135–147

Pikaard CS (1999) Nucleolar dominance and silencing of transcription. Trends Plant Sci 4:478–483

Pikaard CS (2000) The epigenetics of nucleolar dominance. Trends Genet 16:495–500

Pontes O, Neves N, Silva M, Lewis MS, Madlung A, Comai L, Viegas W, Pikaard CS (2004) Chromosomal locus rearrangements are a rapid response to formation of the allotetraploid *Arabidopsis suecica* genome. Proc Natl Acad Sci U S A 101:8240–8245

Raina SN, Parida A, Koul KK, Salimath SS, Bisht MS, Raja V, Khoshoo TN (1994) Associated chromosomal DNA changes in polyploids. Genome 37:560–564

Rapp RA, Wendel J (2005) Epigenetics and plant evolution. New Phytol 168:81–91

Rossi V, Motto M, Pelligrini L (1997) Analysis of the methylation pattern of the maize Opaque-2 (O2) promoter and in vitro binding studies indicate that the O2 B-Zip protein and other endosperm factors can bind to methylated target sequences. J Biol Chem 272:13758–13765

Salmon A, Ainouche ML (2010) Polyploidy and DNA methylation: new tools available. Mol Ecol 19:213–215

Salmon A, Ainouche ML, Wendel JF (2005) Genetic and epigenetic consequences of recent hybridization and polyploidy in Spartina (Poaceae). Mol Ecol 14:1163–1175

Salmon A, Clotault J, Jenczewski E, Chable V, Manzanares-Dauleux MJ (2008) *Brassica oleracea* displays a high level of DNA methylation polymorphism. Plant Sci 174:61–70

Shaked H, Kashkush K, Ozkan H, Feldman M, Levy AA (2001) Sequence elimination and cytosine methylation are rapid and reproducible responses of the genome to wide hybridization and allopolyploidy in wheat. Plant Cell 13:1749–1759

Sheehan KB, McInnerney K, Purevdorj-Gage B, Altenburg SD, Hyman LE (2007) Yeast genomic expression patterns in response to low-shear modeled microgravity. BMC Genomics 8:3

Soltis DE, Soltis PS, Pires JC, Kovarik A, Tate J, Mavrodiev E (2004) Recent and recurrent polyploidy in Tragopogon (Asteraceae): cytogenetic, genomic and genetic comparisons. Biol J Linn Soc 82:485–501

Song K, PL L, Tang K, Osborn TC (1995) Rapid genome change in synthetic polyploids of Brassica and its implications for polyploid evolution. Proc Natl Acad Sci U S A 92:7719–7723

Soppe WJJ, Jacobsen SE, Alonso-Blanco C, Jackson JP, Kakutani T, Koornneef M, Peeters AJM (2000) The late flowering phenotype of *fwa* mutants is caused by gain-of-function epigenetic alleles of a homeodomain gene. Mol Cell 6:791–802

Stokes TL, Richards EJ (2002) Induced instability of two *Arabidopsis* constitutive pathogen-response alleles. Proc Natl Acad Sci U S A 99:7792–7796

Storchova Z, Breneman A, Cande J, Dunn J, Burbank K, O'Toole E, Pellman D (2006) Genome-wide genetic analysis of polyploidy in yeast. Nature 443(7111):541–547

Stupar RM, Bhaskar PB, Yandell BS, Rensink WA, Hart AL, Ouyang S, Veilleux RE, Busse JS, Erhardt RJ, Buell CR, Jiang J (2007) Phenotypic and transcriptomic changes associated with potato autopolyploidization. Genetics 176:2055–2067

Tate JA, Ni ZF, Scheen AC, Koh J, Gilbert CA, Lefkowitz D, Chen ZJ, Soltis PS, Soltis DE (2006) Evolution and expression of homeologous loci in *Tragopogon miscellus* (Asteraceae), a recent and reciprocally formed allopolyploid. Genetics 173:1599–1611

Town CD, Cheung F, Maiti R, Crabtree J, Haas BJ, Wortman JR, Hine EE, Althoff R, Arbogast TS, Tallon LJ, Vigouroux M, Trick M, Bancroft I (2006) Comparative genomics of *Brassica oleracea* and *Arabidopsis thaliana* reveal gene loss, fragmentation, and dispersal after polyploidy. Plant Cell 18:1348–1359

Tu Y, Sun J, Ge X, Li Z (2009) Chromosome elimination, addition and introgression in intertribal partial hybrids between *Brassica rapa* and *Isatis indigotica*. Ann Bot 103:1039–1048

Udall JA, Wendel JF (2006) Polyploidy and crop improvement. Crop Sci 46:3–14

Van Damme EJ, Culerrier R, Barre A, Alvarez R, Rouge P et al (2007) A novel family of lectins evolutionarily related to class V chitinases: an example of neofunctionalization in legumes. Plant Physiol 144:662–672

Verhoeven KJF, Dijk PJV, Biere A (2010) Changes in genomic methylation patterns during the formation of triploid asexual dandelion lineages. Mol Ecol 19:315–324

Wang J, Tian L, Madlung A, Lee HS, Chen M, Lee JJ, Watson B, Kagochi T, Comai L, Chen ZJ (2004) Stochastic and epigenetic changes of gene expression in *Arabidopsis* polyploids. Genetics 167:1961–1973

Wang J, Tian L, Lee HS, Wei NE, Jiang H, Watson B, Madlung A, Osborn TC, Doerge RW, Comai L, Chen ZJ (2006) Genome-wide nonadditive gene regulation in *Arabidopsis* allotetraploids. Genetics 172:507–517

Wang Z, Gerstein M, Snyder M (2009) RNA-Seq: a revolutionary tool for transcriptomics. Nat Rev Genet 10:57–63

Wang H, Jiang J, Chen S, Fang W, Guan Z, Liao Y, Chen F (2013) Rapid genomic and transcriptomic alterations induced by wide hybridization: *Chrysanthemum nankingense* × *Tanacetum vulgare* and *C. crassum* × *Crossostephium chinense* (Asteraceae). BMC Genomics 14:902

Wassenegger M (2000) RNA-directed DNA methylation. Plant Mol Biol 43:203–220

Waugh O'Neill RJ, O'Neill MJ, Marshall Graves JA (1998) Undermethylation associated with retroelement activation and chromosome remodelling in an interspecific mammalian hybrid. Nature 393:68–72

Waugh R, McLean K, Flavell AJ, Pearce SR, Kumar A, Thomas BBT, Powell W (1997) Genetic distribution of Bare–1-like retrotransposable elements in the barley genome revealed by sequence-specific amplification polymorphisms (S-SAP). Mol Gen Genet 253:687–694

Weiss-Schneeweiss H, Schneeweiss GM, Stuessy TF, Mabuchi T, Park JM, Jang CG, Sun BY (2007) Chromosomal stasis in diploids contrasts with genome restructuring in auto- and allopolyploid taxa of *Hepatica* (Ranunculaceae). New Phytol 174:669–682

Whittle CA, Krochko JE (2009) Transcript profiling provides evidence of functional divergence and expression networks among ribosomal protein gene paralogs in *Brassica napus*. Plant Cell 21:2203–2219. https://doi.org/10.1105/tpc.109.068411

Wolffe AP, Matzke MA (1999) Epigenetics: regulation through repression. Science 286:481–486

Xiong LZ, CG X, SaghaiMaroof MA, Zhang Q (1999) Patterns of cytosine methylation in an elite rice hybrid and its parental lines, detected by a methylation-sensitive amplification polymorphism technique. Mol Gen Genet 261:439–446

Xiong Z, Gaeta RT, Pires JC (2011) Homoeologous shuffling and chromosome compensation maintain genome balance in resynthesized allopolyploid *Brassica napus*. Proc Natl Acad Sci U S A 108:7908–7913

Xu ML, Li XQ, Korban SS (2000) AFLP-based detection of DNA methylation. Plant Mol Biol Report 18:321–330

Yamaguchi M, Nagai R, Hashimoto R, Ueda R, Taguchi O (2006) In vivo function of Drosophila Syntrophin, a component of the DGC: analysis with transgenic flies for overexpression or knockdown. A Dros Res Conf 47 :835C

Yi H, Bojja RS, Fu J, Du L (2005) Direct evidence for the function of FUM13 in 3ketoreduction of mycotoxin fumonisins in *Fusarium verticillioides*. JAgric. Food Chem 53(13):5456–5460

Zhang L et al (2008) Spn1 regulates the recruitment of Spt6 and the Swi/Snf complex during transcriptional activation by RNA polymerase II. Mol Cell Biol 28(4):1393–1403

Zhang M, Xu C, Yan H, Zhao N, Wettstein DV, Liu B (2009) Limited tissue culture-induced mutations and linked epigenetic modifications in F1 hybrids of sorghum pure lines are accompanied by increased transcription of DNA methyltransferases and 5-methylcytosine glycosylases. Plant J 57:666–679

Zhao XP, Si Y, Hanson RE, Crane CF, Price HJ, Stelly DM, Wendel JF, Paterson AH (1998) Dispersed repetitive DNA has spread to new genomes since polyploid formation in cotton. Genome Res 8:479–492

Zilberman D, Henikoff S (2007) Genome-wide analysis of DNA methylation patterns. Development 134:3959–3965

7 Polyploidy in Changing Environment

7.1 Introduction

In nature, polyploidy is essential for diversity of new species (maintenance and emergence). Moreover, the environment might also play a role in creating polyploids as they have been found more in extreme environments such as high-altitude, subarctic and xeric areas (Love and Love 1949; Love 1953; Hanelt 1966; Grant 1971). It has been proposed that polyploids can establish better than diploids, and various studies on morphology, physiology and development have been conducted to verify the higher stress tolerance/resistance. It has been seen morphologically that polyploids have large cell sizes (Melaragno et al. 1993) and larger stomata (Hodgson et al. 2010). It has been suggested that genome duplication in polyploids might help to predict the water relations in them by showing changes in stomatal pore size, e.g. in *Betula papyrifera* (birch), polyploids have lesser stomata than diploids resulting in decreased gaseous exchange (Li et al. 1996). Furthermore, other morphological factors such as leaf, cuticle thickness and cell wall composition also play a role in maintaining water potential and gaseous exchange (Johnson 1975; Li et al. 1996). Thus, the polyploids being more adaptive than diploids are still inconclusive because of the lack of extreme experimentation. Maherali et al. (2009) studied how genome doubling affected water relations by comparing the natural diploid and tetraploid with neotetraploids (colchicine-induced) of *Chamerion angustifolium* (firewood). They reported that all tetraploids had large size stomata, increased stem and wall diameter and decreased specific hydraulic conductivity than the diploid ones. However, they noted that the natural tetraploids were significantly drought tolerant than diploids and neotetraploids suggesting that tolerance trait did not evolve instantaneously but evolved later on independent of the genome duplication (Maherali et al. 2009). However, Ramsey (2011) performed the transplant experiment in the field by transferring seedlings from greenhouse to dry dune habitat and noted that there were fivefold increases in survivorship of hexaploid seedlings raised from seeds collected from the field compared with the tetraploid seedlings. He also observed that there was 70% increase in survivorship of

T.-Ul.-H. Dar, R.-Ul. Rehman, *Polyploidy: Recent Trends and Future Perspectives*, https://doi.org/10.1007/978-81-322-3772-3_7

spontaneous neohexaploid seedlings. Thus, polyploids adapt better to changing environments. Furthermore, an integrative experimentation using *Arabidopsis* for examining the role of ecophysiological conditions on morphological alterations in polyploids found the cell size variability between diploids and tetraploids (Li et al. 2012).

In light of the few studies conducted, it seems that polyploidization enhanced the survivorship rate of species. Polyploidy has also shown its potential from the evolutionary point of view as in the earth's recorded history, there have been five major extinctions which are predominantly brought about by climatic changes, volcanic eruptions and strikes of giant meteors. However, it is being presumed that the sixth mass extinction is in progress brought about by human activities. It is evidenced that about 65 mya an extinction event called K-T (Cretaceous-Tertiary) took place during which survival of polyploids was more than the diploid progenitors. This extinction event caused by the volcanic activity and asteroid impacts resulted in low light availability for photosynthesis processes and low freezing temperature (Fawcett et al. 2009). It has been reported that even in vertebrates the polyploidization reduced the probability of extinction (Crow and Wagner 2006).

The innate properties of polyploids make them survive in adverse conditions thus making them very successful invaders which can colonize in these areas, e.g. their abundance in arctic region; this successful invasion occurred after glaciations (Brechmann et al. 2004; Ellstrand and Schierenbeck 2000; Pandit et al. 2006). Many of the plant species have become extinct during the drastic climatic changes which occurred 65 mya, but the invasive polyploids due to their survival capacity and reduced competition spread to new habitats rapidly (Prentis et al. 2008). The polyploids have the capacity to colonize wastelands including the industrial wastelands, e.g. the allopolyploids of *Senecio eboracensis* appearing in man-made habitats (Lowe and Abbott 2003). This property of polyploids to colonize new habitats makes them naturally preferred over diploids to withstand drastic climatic changes.

This can be attributed to changes in polyploids at genetic, physiological, morphological and developmental levels helping them to enter the new niches or improving their fitness in drastic environments (Doyle et al. 2008). The polyploids in comparison to diploids are better suited to resist environmental fluxes which might contribute to genetic redundancy and polysomic inheritances. The polyploids adapted in several different ways such as exhibiting increase in phenotypic variability which is a combined effect of homologous recombination, aneuploidy, chromosome substitution and finally epigenetic changes, even though allo- and autopolyploids show more variations owing to phenotypic consequences due to greater numerical chromosome representation and recombination events.

The mechanisms of adaptations in polyploidy by novel variations is not fully clear, however one view mostly held is that the duplicate genes have less contrast, functionality and thus diverse successfully creating new phenotype, this process is called as neo-functionalization. It has been evidenced that doubling of genes leads to new gene functions and differential expression of genes which is tissue specific helping the polyploidy in diversification and evolutionary success (Doyle et al. 2008). The gene expression changes in results in broader phenotypic changes in polyploidy than diploid progenitors and besides having an added advantage of self-fertilization and apomixes for combating adversities especially the sexual mates are already extinct (Barringer 2007). Thus, the processes of neo-and sub-functionalization are said to promote the hybrid vigour (heterosis effect). All these added changes in polyploidy help them in surviving under the drastic unfavourable conditions in comparison to diploids.

7.2 Adaptive Gene Duplications in Changing Environments

7.2.1 Nutrient Transport

The limited nutrient supply/availability that affects the species adaptation is a well-known fact, e.g. yeast hexose transporter; yeast acquires adaptation to limited nutrients (low glucose by gene duplication HXT6 and HXT7) which increases the expression levels of hexose transporters and also increases the glucose movement into the cell (Brown et al. 1998). It has also been shown by the competitive experiments that the strains with gene duplication out competes the parental strains. Kondrashov et al. (2002) said that HXT6 and HXT7 gene duplication are important owing to their relevance/relationship to distant paralogues in the genome suggesting that these gene duplication events help in adapting shifting mechanisms which help in glucose synthesis under stressed conditions or under low-glucose environment. In other experiments, it has been shown that few strains overexpressing the hexose transporters in limiting the glucose conditions and others in populations reacted to low glucose by adapting different strategies or mechanisms (Gresham et al. 2008).

7.2.2 Heat Resistance

In prokaryotes such as *E. coli*, the duplication of genome provided the adaptation against the heat stress (Riehle et al. 2001). Few genes such as gene for heat stress in this duplicated region were highly expressed thus conferring the increased fitness and developmental processes. It has been shown in *E. coli* that the duplication of evgA gene (transcriptional factor) part of the signalling pathway regulating 37 genes results in temperature withstanding over 50 °C (Christ and Chin 2008). Similar observations of the segmental duplication in chromosomes have also been reported in *Arabidopsis* and yeast (James et al. 2008; DeBolt 2010).

7.2.3 Cold Resistance

Similarly, it has been suggested in many studies that the copy number variation (CNV) in the genomes of several organisms of artic regions provided the cold resistance to them, for example, in Antarctic cod in addition to the AFGP gene family, there were about 177 genes which were overexpressed in comparison to warm water cod. The DNA microarray studies have revealed that 118 genes were overexpressed and some of them were duplicated several times, e.g. AFGP, FBP3211 and an F-type lectin, which have been shown to play active role in cold adaptation of this species (Chen et al. 2008).

7.2.4 Optimal Dosage and Fitness Maintenance

Gene interactions are very important for protein turnover and for achieving fitness by an organism, for example, yeasts with genome duplication; the interacting proteins stay together (Papp et al. 2003) and also help in maintaining the optimum dosage balance for its optimum/favourable growth. In yeast, two paralogous histone proteins H2A and H2B were found to be coded by HTA1-HTB1 and HTA2-HTB2, and its fitness depends on the overall dosage maintained by the duplication of HTA2-HTB2 locus (Libuda and Winston 2006). In *Salmonella typhimurium*, the increased resistance to protamine resulted in the mutation of haeme enzyme (hem C) resulting in the declining growth rate (Pranting and Andersson 2011). However, upon amplification of hem C region, there was restoration of fitness to some extent. Similarly, in *Salmonella enteric*, the mutation resulted in loss of fitness causing decreased antibiotic resistance; however, when the copy number of the gene was increased, there was a significant improvement of antibiotic resistance (Paulander et al. 2010).

7.2.5 Salt Protection

In yeast, it has also been observed that gene duplication does provide tolerance to salt stress. There was an increase in several salt responsive genes during salt treatments which provided better resistance. The individual gene duplication events did not provide the same effects (Dhar et al. 2011) which imply that consistent polyploidy is required for efficient adaptation against salt (Dhar et al. 2011; Gerstein et al. 2006). Similar results of different polyploids showing resistance against salt stress have been reported for citrus (Saleh et al. 2008) and sorghum (Ceccarelli et al. 2006) which is indicative of the fact that polyploidy is of physiological significance which provides and adaptive response against osmotic stress (Dhar et al. 2011).

7.2.6 Heavy Metal Resistance

The polyploidy-induced metal resistance in many species might be linked to cationic movement outside the cell (Kondrashov et al. 2002, Kondrashov and Kondrashov 2006). A study involving bacteria *Ralstonia pickettii* growing in contaminated lake showed an accumulation of high concentration of copper in their outer membrane, and their genome analysis revealed the duplication of the genomic regions containing metal resistance and their transporter genes (Yang et al. 2010). Similarly, the genome analysis of the bacteria *Cupriavidus metallidurans* revealed that an increase in the copy number of genes by duplication was responsible for the uptake of metals from the surrounding environment (von Rozycki and Nies 2009).

7.2.7 Antibiotic and Drug Resistance

The antibiotic resistance in bacteria is due to polyploidy of certain genes (Kondrashov and Kondrashov 2006; Craven et al. 2007; Sandegren and Andersson 2009). In the similar way, the response of tumour cells to various drug treatments is of major concern due to duplications or gene amplifications (Kondrashov and Kondrashov 2006). The amplification of genes upon drug treatments is not observed only in microbes and somatic cells but also observed upon treating malaria (Templeton 2009) and *Leishmania* (Segovia 1994). The example of multidrug resistance gene of *Plasmodium falciparum* is pfmdr, which has evolved over time and attained resistance due to widespread use of chloroquine of *Cinchona* tree. Due to the frequent use of chloroquine, the pfmdr1 gene is increasingly amplified resulting in resistance against other antimalarial drug treatments also. However, evidences also point out that pfmdr1 gene has undergone duplication independently many times in nature (Duraisingh and Cowman 2005; Preechapornkul et al. 2009; Chavchich et al. 2010; Triglia e al. 1991). Similarly, the gene amplification in *Leishmania* has been observed as an adaptation to drugs (Segovia 1994), and it has commonly been found in its circular extrachromosomal DNA elements. The bacterial strain *E. coli* also behaved in similar fashion by restoring to gene amplification causing the antibiotic resistance in toxic environments thus leading to their increased fitness (Soo et al. 2011). Therefore, gene amplification process in an organism in response to antibiotics or drugs in an environment helps it in improving their overall adaptative capacity.

7.2.8 Resistance to Pesticides and Complex Organic Compounds

It has been seen that the pesticides or organic compounds application if prolonged leads to the duplication of genes which are responsible for production of those enzymes which degrade these chemicals, thus rendering the organism resistant towards it (Taylor and Feyereisen 1996; Bass and Field 2011). For example, when the mosquito *Culex pipiens* were subjected to the treatments of pesticides, it leads to the duplication of two genes encoding non-specific esterase enzymes (Est-2 and Est-3) and acetylcholine esterase (Ace.1) (Kondrashov and Kondrashov 2006; Labbé et al. 2007). The Ace.1 is a target of organophosphate insecticides and the mosquito has evolved Ace.1 locus, enabling populations to have duplications predominantly in heterozygous state as the duplications in homozygous state are not significant and do not help an organism for fitness. This persistence of Ace.1 duplication in heterozygous state indicates the positive selection.

7.3 Adaptation to Domestication

In various breeds of cow, variability has been observed in copy number (CP), e.g. the copy number variations (CNVs) of genes related to immunity especially the genes encoding UL16 binding proteins when analysed in different breeds of cattle have revealed that only fixed copies of these genes have been duplicated in the ancestors of Bos lineage. Further, some of the duplicated genes have been seen to be associated with phenotypes related to domestication such as proteins involved for milk production and ruminant digestion (Liu et al. 2009). It has also been pointed out that breed-specific phenotypes could be related to gene function which is indicative of the fact that health and productivity-related aspects could also be selected by copy number variations (CNVs) during the domestication process.

7.4 Gene Duplications in Humans

Several examples are there when gene duplication confers resistance towards diseases to provide fitness advantage, e.g. the duplication of CCL31 gene (HIV suppressive CC chemokine receptor) results in lower susceptibility towards HIV infection, thus delaying the development of AIDS after getting infected (Gonzalez et al. 2005). Also, the higher copy number of AMY1 (salivary amylase gene) will result in more salivary amylase that ascribes higher fitness value to them (Perry et al. 2007). Many more such examples might be there in the human genome carrying higher significance which is yet to be described (Lupski 2007; Cooper et al. 2007; Emerson et al. 2008; Gazave et al. 2011).

7.5 General View Regarding Adaptive Duplication

There are innumerable examples of gene duplication resulting in adaptation of different species towards different environmental conditions. It has been indicated that duplication of genes are not rare mechanisms, but they are positively associated with the adaptability of organisms in its environment. The protein coded by the duplicated genes is as good as the coded by the normal gene, and sometimes they perform even better upon interaction with toxins, nutrients and foreign bodies in varying environments. As gene duplication is an adaptive phenomenon, mostly the gene products should be produced at much higher concentration to meet the demand, for example, storage proteins and antifreeze proteins. It has been hypothesized that the role played by gene duplication for adaptive purposes is totally different from fixed gene copies in functional genome (Velkov 1982; Romero and Palacios, 1997; Kondrashov et al. 2002; Hastings 2007; Flagel and Wendel 2009; Kondrashov and Kondrashov 2006). Moreover, it is very difficult to point out the accurate list of genes for a particular resistance during certain environmental challenge based on the functionality and the gene duplication in the genome. For example, the duplication of the genes in olfactory receptor protein responsible for increased sensitivity towards particular odour provides adaptation under certain conditions (Nei et al. 2008; Niimura 2009; Lavagnino et al. 2012). Furthermore, opsin gene duplication results in adaptation in sensitivity to light of certain wavelengths (Sison-Mangus et al. 2006; Gojobori and Innan 2009). The duplication in pathogenicity (Powell et al. 2008; Jiang et al. 2010) and pathogen resistance (Hanada et al. 2008; Cornman and Willis 2008; Gong et al. 2010; Schulenburg and Boehnisch 2008; Neiman et al. 2009) genes suggests that they are involved in co-evolution of parasite and its host or any other symbiotic relationship. It has also been observed that gene duplication related to osmotic stresses is advantageous in various plant species resistant to drought. The duplication of genes is advantageous in extreme conditions such as the high altitude, as in lower oxygen concentrations, it would be helpful for the optimization of oxygen transport. Duplication of genes are also effective in regulating the toxin dosage whose effectiveness is regulated by gene duplication of toxins and venoms (Hoffmann et al. 2008; Storz et al. 2009; Wong and Belov 2012; Duda and Palumbi 1999; Moran and Gurevitz 2006; Lynch 2007; Murray et al. 2011; Ma et al. 2012; Chang and Duda Chang and Duda 2012). Duplication of genes has been shown to be involved in neutralization of pyrrolizidine alkaloids by synthesizing proteins required for degradation and thus conferring adaptive value to a herbivore feeding on plants having hyper-concentration of specific toxic compounds (Wang et al. 2012). Similarly, gene duplication in developmental processes in organisms may provide adaptation by modifying certain physiological and anatomical features for the new environment. It is being investigated that luciferase phenomenon and the carotenoid biosynthesis might be controlled by gene duplication leading to increased production of pigment of light emission.

References

Barringer BC (2007) Polyploidy and self-fertilization in flowering plants. Am J Bot 94:1527–1533

Bass C, Field LM (2011) Gene amplification and insecticide resistance. Pest Manag Sci 67:886–890

Brechmann C, Brysting AK, Alsos IG, Borgen L, Grundt HH, Scheen AC, Elven R (2004) Polyploidy in arctic plants. Biol J Linn Soc 82:521–536

Brown CJ, Todd KM, Rosenzweig RF (1998) Multiple duplications of yeast hexose transport genes in response to selection in a glucose-limited environment. Mol Biol Evol 15:931–942. https://doi.org/10.1093/oxfordjournals.molbev.a026009

Ceccarelli M, Santantonio E, Marmottini F, Amzallag GN, Cionini PG (2006) Chromosome endoreduplication as a factor of salt adaptation in *Sorghum bicolor*. Protoplasma 227:113–118. https://doi.org/10.1007/s00709-005-0144-0

Chang D, Duda TF Jr (2012) Extensive and continuous duplication facilitates rapid evolution and diversification of gene families. Mol Biol Evol 29:2019–2029

Chavchich M, Gerena L, Peters J, Chen N, Cheng Q, Kyle DE (2010) Role of pfmdr1 amplification and expression in induction of resistance to artemisinin derivatives in *Plasmodium falciparum*. Antimicrob Agents Chemother 54:2455–2464

Chen Z et al (2008) Transcriptomic and genomic evolution under constant cold in Antarctic notothenioid fish. Proc Natl Acad Sci U S A 105(12):944–949. https://doi.org/10.1073/pnas.0802432105

Christ D, Chin JW (2008) Engineering Escherichia coli heat-resistance by synthetic gene amplification. Protein Eng Des Sel 21:121–125. https://doi.org/10.1093/protein/gzm085

Cooper GM, Nickerson DA, Eichler EE (2007) Mutational and selective effects on copy-number variants in the human genome. Nat Genet 39:S22–S29

Cornman RS, Willis JH (2008) Extensive gene amplification and concerted evolution within the CPR family of cuticular proteins in mosquitoes. Insect Biochem Mol Biol 38:661–676

Crow KD, Wagner GP (2006) What is the role of genome duplication in the evolution of complexity and diversity? Mol Biol Evol 23:887–892

DeBolt S (2010) Copy number variation shapes genome diversity in Arabidopsis over immediate family generational scales. Genome Biol Evol 2:441–453. https://doi.org/10.1093/gbe/evq033

Dhar R, Sagesser R, Weikert C, Yuan J, Wagner A (2011) Adaptation of *Saccharomyces cerevisiae* to saline stress through laboratory evolution. J Evol Biol 24:1135–1153. https://doi.org/10.1111/j.1420-9101.2011.02249.x

Doyle JJ, Flagel LE, Paterson AH, Rapp RA, Soltis DE, Soltis PS, Wendel JF (2008) Evolutionary genetics of genome merger and doubling in plants. Annu Rev Gen 42:443–461

DudaJr TF, Palumbi SR (1999) Molecular genetics of ecological diversification: duplication and rapid evolution of toxin genes of the venomous gastropod Conus. Proc Natl Acad Sci U S A 96:6820–6823

Duraisingh MT, Cowman AF (2005) Contribution of the pfmdr1 gene to antimalarial drug-resistance. Acta Trop 94:181–190

Ellstrand NC, Schierenbeck KA (2000) Hybridization as a stimulus for the evolution of invasiveness in plants? Proc Natl Acad Sci USA 97:7043–7050

Emerson JJ, Cardoso-Moreira M, Borevitz JO, Long M (2008) Natural selection shapes genome-wide patterns of copy-number polymorphism in *Drosophila melanogaster*. Science 320:1629–1631

Fawcett JA, Maerea S, de Peera YV (2009) Plants with double genomes might have had a better chance to survive the Cretaceous–Tertiary extinction event. Proc Natl Acad Sci USA 106:5737–5743

Flagel LE, Wendel JF (2009) Gene duplication and evolutionary novelty in plants. New Phytol 183:557–564

Gazave E et al (2011) Copy number variation analysis in the great apes reveals species-specific patterns of structural variation. Genome Res 21:1626–1639

Gerstein AC, Chun HJ, Grant A, Otto SP (2006) Genomic convergence toward diploidy in *Saccharomyces cerevisiae*. PLoS Genet 2:e145. https://doi.org/10.1371/journal. pgen.0020145

Gojobori J, Innan H (2009) Potential of fish opsin gene duplications to evolve new adaptive functions. Trends Genet 25:198–202

Gong D, Wilson PW, Bain MM, McDade K, Kalina J, Herve´-Gre´pinet V, Nys Y, Dunn IC (2010) Gallin; an antimicrobial peptide member of anew avian defensin family, the ovodefensins, has been subject to recent gene duplication. BMC Immunol 11:12

Gonzalez E et al (2005) The influence of CCL3L1 gene-containing segmental duplications on HIV-1/AIDS susceptibility. Science 307:1434–1440

Grant V (1971) Plant speciation, 1st edn. Columbia University Press, New York

Gresham D, Desai MM, Tucker CM, Jenq HT, Pai DA, Ward A, DeSevo CG, Botstein D, Dunham MJ (2008) The repertoire and dynamics of evolutionary adaptations to controlled nutrient-limited environments in yeast. PLoS Genet 4:e1000303. https://doi.org/10.1371/journal.pgen.1000303

Hanada K, Zou C, Lehti-Shiu MD, Shinozaki K, Shiu SH (2008) Importance of lineage-specific expansion of plant tandem duplicates in the adaptive response to environmental stimuli. Plant Physiol 148:993–1003

Hanelt P (1966) Polyploidie-frequenz und geographischeverbreitungbeihoherenpflanzen. BiologischeRundschau 4:183–196

Hastings PJ (2007) Adaptive amplification. Crit Rev Biochem Mol Biol 42:271–283

Hodgson JG, Sharafi M, Jalili A, Diaz S, Montserrat-Marti G, Palmer C et al (2010) Stomatal vs. genome size in angiosperms: the somatic tail wagging the genomic dog? Ann Bot (Lond) 105:573–584

Hoffmann FG, Opazo JC, Storz JF (2008) Rapid rates of lineage-specific gene duplication and deletionin the alpha-globin gene family. Mol Biol Evol 25:591–602

James TC, Usher J, Campbell S, Bond U (2008) Lager yeasts possess dynamic genomes that undergo rearrangements and gene amplification in response to stress. Curr Genet 53:139–152. https://doi.org/10.1007/s00294-007-0172-8

Jiang SY, Ma Z, Ramachandran S (2010) Evolutionary history and stress regulation of the lectin superfamily in higher plants. BMC Evol Biol 10:79

Johnson HB (1975) Plant pubescence: an ecological perspective. Bot Rev 41:233–258

Kondrashov FA, Kondrashov AS (2006) Role of selection in fixation of gene duplications. J Theor Biol 239:141–151. https://doi.org/10.1016/j.jtbi.2005.08.033

Kondrashov FA, Rogozin IB, Wolf YI, Koonin EV (2002) Selection in the evolution of gene duplications. Genome Biol 3, research0008. https://doi.org/10.1186/gb-2002-3-2-research0008

Labbé P, Berticat C, Berthomieu A, Unal S, Bernard C, Weill M, Lenormand T (2007) Forty years of erratic insecticide resistance evolution in the mosquito *Culex pipiens*. PLoS Genet 3:e205

Lavagnino N, Serra F, Arbiza L, Dopazo H, Hasson E (2012) Evolutionary genomics of genes involved in olfactory behavior in the *Drosophila melanogaster* species group. Evol Bioinformatics Online 8:89–104

Levin D (2002) The role of chromosomal change in plant evolution. Oxford University Press, New York

Li W-L, Berlyn GP, Ashton PMS (1996) Polyploids and their structural and physiological characteristics relative to water deficit in *Betula papyrifera* (Betulaceae). Am J Bot 83:15–20

Li X, Yu E, Fan C, Zhang C, Fu T, Zhou Y (2012) Developmental, cytological and transcriptional analysis of autotetraploid Arabidopsis. Planta 236:579–596

Libuda DE, Winston F (2006) Amplification of histone genes by circular chromosome formation in *Saccharomyces cerevisiae*. Nature 443:1003–1007. https://doi.org/10.1038/nature05205

Liu GE, Ventura M, Cellamare A, Chen L, Cheng Z, Zhu B, Li C, Song J, Eichler EE (2009) Analysis of recent segmental duplications in the bovine genome. BMC Genomics 10:571

Love A (1953) Subarctic polyploidy. Hereditas 39:113–124

Love A, Love D (1949) The geobotanical significance of polyploidy. Portugaliae Acta (Suppl):273–352

Lowe AJ, Abbott RJ (2003) A new British species, Senecioeboracensis(Asteraceae), another hybrid derivative of S. vulgaris L. and S. squalidus L. Watsonia 24:375–388

Lupski JR (2007) An evolution revolution provides further revelation. BioEssays 29:1182–1184

Lynch VJ (2007) Inventing an arsenal: adaptive evolution and neofunctionalization of snake venom phospholipase A2 genes. BMC Evol Biol 7:2

Ma Y, He Y, Zhao R, Wu Y, Li W, Cao Z (2012) Extreme diversity of scorpion venom peptides and proteins revealed by transcriptomic analysis: implication for proteome evolution of scorpion venom arsenal. J Proteome 75:1563–1576

Maherali H, Walden AE, Husband BC (2009) Genome duplication and the evolution of physiological responses to water stress. New Phytol 184:721–731

Melaragno JE, Mehrotra B, Coleman AW (1993) Relationship between endopolyploidy and cell size in epidermal tissue of arabidopsis. Plant Cell Online 5:1661–1668

Moran Y, Gurevitz M (2006) When positive selection of neurotoxin genes is missing. The riddle of the sea anemone *Nematostella vectensis*. FEBS J 273:3886–3892

Murray SA, Mihali TK, Neilan BA (2011) Extraordinary conservation, gene loss, and positive selection in the evolution of an ancient neurotoxin. Mol Biol Evol 28:1173–1182

Nei M, Niimura Y, Nozawa M (2008) The evolution of animal chemosensory receptor gene repertoires: roles of chance and necessity. Nat Rev Genet 9:951–963

Neiman M, Olson MS, Tiffin P (2009) Selective histories of poplar protease inhibitors: elevated polymorphism, purifying selection, and positive selection driving divergence of recent duplicates. New Phytol 183:740–750

Niimura Y (2009) Evolutionary dynamics of olfactory receptor genes in chordates: interaction between environments and genomic contents. Hum Genomics 4:107–118

Pandit MK, Tan HTW, Bisht MS (2006) Polyploidy in invasive plant species of Singapore. Bot J Linn Soc 151:395–403

Papp B, Pa'l C, Hurst LD (2003) Dosage sensitivity and the evolution of gene families in yeast. Nature 424:194–197. https://doi.org/10.1038/nature01771

Paulander W, Andersson DI, Maisnier-Patin S (2010) Amplification of the gene for isoleucyl-tRNAsynthetase facilitates adaptation to the fitness cost of mupirocin resistance in *Salmonella enterica*. Genetics 185:305–312. https://doi.org/10.1534/genetics.109.113514

Perry GH et al (2007) Diet and the evolution of human amylase gene copy number variation. Nat Genet 39:1256–1260

Powell AJ, Conant GC, Brown DE, Carbone I, Dean RA (2008) Altered patterns of gene duplication and differential gene gain and loss in fungal pathogens. BMC Genomics 9:147

Pranting M, Andersson DI (2011) Escape from growth restriction in small colony variants of *Salmonella typhimurium* by gene amplification and mutation. Mol Microbiol 79:305–315. https://doi.org/10.1111/j.1365-2958. 2010.07458.x

Preechapornkul P, Imwong M, Chotivanich K, PongtavornpinyoW DAM, Day NP, White NJ, Pukrittayakamee S (2009) *Plasmodium falciparum* pfmdr1 amplification, mefloquine resistance, and parasite fitness. Antimicrob Agents Chemother 53:1509–1515

Ramsey J (2011) Polyploidy and ecological adaptation in wild yarrow. Proc Natl Acad Sci 108:7096–7101

Riehle MM, Bennett AF, Long AD (2001) Genetic architecture of thermal adaptation in Escherichia coli. Proc Natl Acad Sci U S A 98:525–530. https://doi.org/10.1073/pnas.98.2.525

Romero D, Palacios R (1997) Gene amplification and genomic plasticity in prokaryotes. Annu Rev Genet 31:91–111

Saleh B, Allario T, Dambier D, Ollitrault P, Morillon R (2008) Tetraploid citrus rootstocks are more tolerant to salt stress than diploid. C R Biol 331:703–710. https://doi.org/10.1016/j.crvi.2008.06.007

Sandegren L, Andersson DI (2009) Bacterial gene amplification: implications for the evolution of antibiotic resistance. Nat Rev Microbiol 7:578–588

Schulenburg H, Boehnisch C (2008) Diversification and adaptive sequence evolution of *Caenorhabditis lysozymes* (Nematoda: Rhabditidae). BMC Evol Biol 8:114

Segovia M (1994) Leishmania gene amplification: a mechanism of drug resistance. Ann Trop Med Parasitol 88:123–130

Sison-Mangus MP, Bernard GD, Lampel J, Briscoe AD (2006) Beauty in the eye of the beholder: the two blue opsins of lycaenid butterflies and theopsin gene-driven evolution of sexually dimorphic eyes. J Exp Biol 209:3079–3090

Soo VW, Hanson-Manful P, Patrick WM (2011) Artificial gene amplification reveals an abundance of promiscuous resistance determinants in *Escherichia coli*. Proc Natl Acad Sci USA 108:1484–1489

Storz JF, Runck AM, Sabatino SJ, Kelly JK, Ferrand N, Moriyama H, Weber RE, Fago A (2009) Evolutionary and functional insights into the mechanism underlying high-altitude adaptation of deer mouse hemoglobin. Proc Natl Acad Sci U S A 106:14450–14455

Taylor M, Feyereisen R (1996) Molecular biology and evolution of resistance of toxicants. Mol Biol Evol 13:719–734

Templeton TJ (2009) The varieties of gene amplification, diversification and hypervariability in the human malarial parasite, *Plasmodium falciparum*. Mol Biochem Parasitol 166:109–116

Triglia T, Foote SJ, Kemp DJ, Cowman AF (1991) Amplification of the multidrug resistance gene pfmdr1 in *Plasmodium falciparum* has arisen as multiple independent events. Mol Cell Biol 11:5244–5250

Velkov VV (1982) Gene amplification in prokaryotic and eukaryotic systems. Genetika 18:529–543

von Rozycki T, Nies DH (2009) *Cupriavidus metallidurans*: evolution of a metal-resistant bacterium. Antonie Van Leeuwenhoek 96:115–139

Wang L, Beuerle T, Timbilla J, Ober D (2012) Independent recruitment of a flavin-dependent monooxygenase for safe accumulation of sequestered pyrrolizidine alkaloids in grasshoppers and moths. PLoS One 7:e31796

Wong ES, Belov K (2012) Venom evolution through gene duplications. Gene 496:1–7

Yang F, Pecina DA, Kelly SD, Kim SH, Kemner KM, Long DT, Marsh TL (2010) Biosequestration via cooperative binding of copper by Ralstonia pickettii. Environ Technol 31:1045–1060

8 Future Prospects in Polyploidy Research

Polyploidy has played a great role in the evolution of domesticated plants (Hilu 1993); indeed, most of the major crop plants are allopolyploids, including wheat (*Triticum aestivum*), maize (*Zea mays*), oats (*Avena sativa*), alfalfa (*Medicago sativa*), sugarcane (*Saccharum officinale*), potato (*Solanum tuberosum*), sweet potato (*Ipomoea batatas*), banana (*Musa sapientum*), cotton (*Gossypium hirsutum*), tobacco (*Nicotiana tabacum*), coffee (*Coffea arabica*), apple (*Malus pumila*) and pear (*Pyrus communis*). With this background, polyploidy has a bright future in such departments as agriculture, horticulture and floriculture. It has great scope of being used in plant breeding to improve the existing crops or develop new ones to cater to the needs of exponentially increasing human population. According to Stebbins (1950), allopolyploids or amphiploids can be useful in plant breeding in three ways:

To transfer a valuable character carried by a single genetic factor across a barrier of interspecific sterility.
To incorporate some desirable characteristic controlled by many genetic factors into a crop plant.
To be used as entirely new crop plants with qualities different from any one known.

Polyploid crops afford many commercial benefits. They are often associated with such characteristics as it would make them useful and beneficial; these include bigger fruits and seeds, seedless fruits (e.g. seedless banana, watermelons and grapes) and disease resistance (e.g. tobacco allotetraploids are resistant to mosaic virus infection). Given the usefulness of polyploids for agriculture, polyploidy has regained the interest of researchers during the first decade of twenty-first century (Chen et al. 2007), with new efforts on *Brassica* spp. (Lukens et al. 2006; Gaeta et al. 2007), cotton (Jiang et al. 1998; Liu et al. 2001), *Nicotiana* spp. (Lim et al. 2006; Pearse et al. 2006; Anssour et al., 2009) and wheat (Han et al. 2003; Clarkson et al. 2004). It is being endeavoured to enhance the commercial benefits of polyploidy in crop plants. Recent advances to clone genes are used in the genomic

T.-Ul.-H. Dar, R.-Ul. Rehman, *Polyploidy: Recent Trends and Future Perspectives*, https://doi.org/10.1007/978-81-322-3772-3_8

research of major crops (Gupta et al. 2008). Polyploidy can be instrumental in our efforts to go for 'perennial grains' and help reach the higher goal of 'agriculture next revolution'.

Floriculture is fast growing as an important commercial business and increasingly new flower types are commonly used as ornamentals. Increasing the flower size and prolonging the flowering period would be desirable traits in such plants. The demand for autotetraploids and larger and variegated hybrid flowers may increase day by day.

Polyploids are known for their tolerance of extreme environmental conditions. As suggested by Fawcett et al. (2009), polyploidy ameliorated extinction risks during the K-T mass extinction. In view of this adaptive hypothesis, it can be guessed that polyploidy may confer a similar advantage in the current period of global warming and in the foreseen sixth mass extinction event. The hottest new trend in climate change may be global cooling, as warned by some researchers (58th Bilderberg Meeting held on 3–6 June 2010 at Sitges in Spain). Their increasing levels of cold tolerance, preferential perennial herbaceous habit, and reproductive capacity through apomixis may be construed to help polyploids thrive in such conditions and contribute significantly to the future world flora! Based on factors that favour establishment of polyploids, it can be envisaged that polyploidy would be more frequent in perennial herbs than in annuals and woody plants; also among perennials, those species with effective means of vegetative reproduction are better adapted for polyploidy. Overgrazing is one of the causes of loss of biodiversity. Since polyploids are known as best colonizers of barren lands, they can be used for regeneration of depleted grazing lands (Stebbins 1950).

Medicinal and aromatic plants (MAPs) are of considerable socio-economic importance in the contemporary world. Since polyploidy can increase various cell substances, it can be used to increase contents of active principles in MAPs and raise their production rate of medicines; this indirectly will lower the pressure on wild-growing MAPs and facilitate their conservation. Polyploidy can also facilitate fixation of heterozygosity in obligate asexual species of medicinal and aromatic plants, which is vital for maintaining uniformity in quality of the economic product and quantity of biomass under seed cultivation (Lavania et al. 2012).

An overall perusal of the important plant genomic literature made available through studies conducted during the last several decades has elucidated that polyploidy may lead to important morphological, developmental, physiological, biochemical, genetic and ecological consequences (Levin 1983). It is these consequences that are considered to cause success of polyploids in nature (e.g. see Tate et al. 2005; Adams 2007; Otto 2007; Doyle et al. 2008; Leitch and Leitch 2008). Soltis and Burleigh (2009) specify that these changes include sub-functionalization, new gene functions, epigenetic changes and altered gene expressions, besides impact on CO_2 exchange rates, hormone levels, photosynthetic rates and water balance. Thompson et al. (2004) indicated that plant polyploidy can have profound effects on mating systems and interactions with animal behaviours and pollinators. Nonetheless, these changes may vary in different groups and differ as per the mode of polyploidization (Tate et al. 2005; Doyle et al. 2008). In view of these facts, Soltis and Burleigh

(2009) caution to be careful in attributing too much evolutionary power and adaptive significance to polyploidy. This has also been earlier noted by Stebbins (1950) that polyploids are complex and generalizations are difficult to make.

The technological upgradations especially in genomics tools have helped in renaissance of polyploidy research. The multidisciplinary approaches have helped comprehensively in elucidating the generated knowledge into practical applications. The application of high-throughput genome-wide sequencing approaches include studying genetic and epigenetic changes of polyploidization as well as the availability of phenomic platforms for phenotyping. These technologies and platforms have become more accessible (due to affordable costs) and also feasible for studying more diverse and large genome species (Grover et al. 2012; Elshire et al. 2011). These technologies will also make possible to discover new genes and markers and also help in genome-wide gene expression. The use of ChIP bisulphite sequencing allows investigation of chromatin and DNA methylation patterns. The high-throughput sequencing of RNA transcripts provides an advantage by circumventing the problems posed by redundant and large genomes. However, the rapid speed of sequencing genomes has made available the large amount of data thus creating disparity between data generation and biologically meaningful analysis. But there are some studies suggesting the strategies for addressing these gaps (Ilut et al. 2012; Grover et al. 2012; Cronn et al. 2012; Kvam et al. 2012). Employing novel statistical models in genetic mapping will help in gene identification during polyploidization process. Combining the results from genetics and omics platforms shall greatly help in understanding the polyploid genome flexibility and comparing it with amplified responses. For understanding the evolution of polyploids, it is required to integrate the knowledge at the population level, designing suitable experiments and survey of variations at various levels. There is a wealth of knowledge on polyploidization (formation, maintenance and divergence) at the whole genome and subgenome levels which would help the plant biologists in understanding evolution and diversification and also help the plant breeders for designing appropriate strategies for crop improvement available (Schmutz et al. 2010; Potato Genome Sequencing Consortium 2012).

References

Adams KL (2007) Evolution of the duplicate gene expression in the polyploidy and hybrid plants. J Hered 98:136–141

Anssour S, Krügel T, Sharbel TF, Saluz HP, Bonaventure G, Baldwin IT (2009) Phenotypic, genetic and genomic consequences of natural and synthetic polyploidization of *Nicotiana attenuata* and *Nicotiana obtusifolia*. Ann Bot 103:1207–1217

Chen ZJ, Scheffler BE, Dennis E, Triplett BA, Zhang T, Guo W, Chen X, Stelly DM, Rabinowicz PD, Town CD, Arioli T (2007) Toward sequencing cotton (Gossypium) genomes. Plant Physiol 145:1303–1310

Clarkson JJ, Knapp S, Garcia VF, Olmstead RG, Leitch AR, Chase MW (2004) Phylogenetic relationships in *Nicotiana* (Solanaceae) inferred from multiple plastid DNA regions. Mol Phylogenet Evol 33:75–90

Cronn R, Knaus BJ, Liston A, Maughan PJ, Parks M, Syring JV, Udall J (2012) Targeted enrichment strategies for next-generation plant biology. Am J Bot 99:291–311

Doyle JJ, Flagel LE, Paterson AH, Rapp RA, Soltis DE, Soltis PS, Wendel JF (2008) Evolutionary genetics of genome merger and doubling in plants. Annu Rev Genet 42:443–461

Elshire RJ, Glaubitz JC, Sun Q, Poland JA, Kawamoto K, Buckler ES, Mitchell SE (2011) A robust, simple genotyping-by-sequencing (GBS) approach for high diversity species. PLoS One 6:e19379

Fawcett JA, Maere S, van de peer Y (2009) Plants with double genomes might have had a better chance to survive the cretaceous-tertiary extinction event. Proc Nat Acad Sci U S A 106:5737–5742

Gaeta RT, Pires JC, Iniguez-Luy F, Leon E, Osborn TC (2007) Genomic changes in resynthesized Brassica napus and their effect on gene expression and phenotype. Plant Cell 19:3403–3417

Grover CE, Salmon A, Wendel JF (2012) Targeted sequence capture as a powerful tool for evolutionary analysis. Am J Bot 99:312–319

Gupta PK, Rustgi S, Mir R (2008) Array-based high-throughput DNA markers for crop improvement. Heredity 101:5–18

Han FP, Fedak G, Ouellet T, Liu B (2003) Rapid genomic changes in interspecific and intergeneric hybrids and allopolyploids of *Triticeae*. Genome 46:716–723

Hilu KW (1993) Polyploidy and the evolution of domesticated plants. Am J Bot 80:1494–1499

Ilut DC, Coate JE, Luciano AK, Owens TG, May GD, Farmer A, Doyle JJ (2012) A comparative transcriptomic study of an allotetraploid and its diploid progenitors illustrates the unique advantages and challenges of RNA-seq in plant species. Am J Bot 99:383–396

Jiang CX, Wright RJ, El-Zik K, Paterson AH (1998) Polyploid formation created unique avenues for response to selection in Gossypium (cotton). Proc Natl Acad Sci U S A 95:4419–4424

Kvam VM, Liu P, Si Y (2012) A comparison of statistical methods for detecting differentially expressed genes from RNA-seq data. Am J Bot 99:248–256

Lavania UC, Srivastava S, Lavania S, Basu S, Misra NK, Mukai Y (2012) Autopolyploidy differentially influences body size in plants, but facilitates enhanced accumulation of secondary metabolites, causing increased cytosine methylation. Plant J 71:539–549

Leitch AR, Leitch IJ (2008) Genomic plasticity and the diversity of polyploid plants. Science 320:481–483

Levin DA (1983) Polyploidy and novelty in flowering plants. Am Nat 122:1–25

Lim KY, Souckova-Skalicka K, Sarasan V, Clarkson JJ, Chase MW, Kovarik A, Leitch AR (2006) A genetic appraisal of a new synthetic *Nicotiana tabacum* (Solanaceae) and the Kostoff synthetic tobacco. Am J Bot 93:875–883

Liu B, Brubaker CL, Mergeai G, Cronn RC, Wendel JF (2001) Polyploid formation in cotton is not accompanied by rapid genomic changes. Genome 44:321–330

Lukens LN, Pires JC, Leon E, Vogelzang R, Oslach L, Osborn T (2006) Patterns of sequence loss and cytosine methylation within a population of newly resynthesized *Brassica napus* allopolyploids. Plant Physiol 140:336–348

Otto SP (2007) The evolutionary consequences of polyploidy. Cell 131:452–462

Pearse IS, Krugel T, Baldwin IT (2006) innovation in anti-herbivore defense systems during neopolypoloidy—the functional consequences of instantaneous speciation. Plant J 47:196–210

Potato Genome Sequencing Consortium (2012) Genome sequence and analysis of the tuber crop potato. Nature 475:189–195

Schmutz J, Cannon SB, Schlueter J, Ma J, Mitros T, Nelson W, Hyten DL, Song Q, Thelen JJ, Cheng J et al (2010) Genome sequence of the palaeopolyploid soybean. Nature 463:178–183

Soltis DE, Burleigh JG (2009) Surviving the K-T mass extinction: new perspectives of polyploidization in angiosperms. Proc Natl Acad Sci U S A 106:5455–5456

Stebbins GL (1950) Variation and evolution in plants. Columbia University Press, New York

Tate JA, Soltis DE, Soltis PS (2005) Polyploidy in plants. In: Gregory TR (ed) The evolution of the genome. Elsevier, San Diego, pp 371–426

Thompson JN, Nuismer SL, Merg KF (2004) Plant polyploidy and the evolutionary ecology of plant/animal interactions. Biol J Linn Soc 82:511–519